Algorithms
for Chemical Computations

Ralph E. Christoffersen, EDITOR

The University of Kansas

A symposium sponsored by

the Division of Computers

in Chemistry at the 171st

Meeting of the American

Chemical Society, New York, N.Y.,

Aug. 30, 1976.

A C S   S Y M P O S I U M   S E R I E S   **46**

AMERICAN CHEMICAL SOCIETY

WASHINGTON, D. C.   1977

Library of Congress CIP Data

Algorithms for chemical computations.
(ACS symposium series; 46 ISSN 0097-6156)
  Includes bibliographical references and index.
  1. Chemistry—Data processing—Congresses. 2. Algo-
rithms—Congresses.
  I. Christoffersen, Ralph E., 1937–   . II. American
Chemical Society. Division of Computers in Chemistry.
III. Series: American Chemical Society. ACS symposium
series; 46.

QD39.3.E46A43   540′.28′5   77-5030
ISBN 0-8412-0371-7   ACSMC8 46 1–151

ACS Symposium Series

Robert F. Gould, *Editor*

FOREWORD

The ACS SYMPOSIUM SERIES was founded in 1974 to provide a medium for publishing symposia quickly in book form. The format of the SERIES parallels that of the continuing ADVANCES IN CHEMISTRY SERIES except that in order to save time the papers are not typeset but are reproduced as they are submitted by the authors in camera-ready form. As a further means of saving time, the papers are not edited or reviewed except by the symposium chairman, who becomes editor of the book. Papers published in the ACS SYMPOSIUM SERIES are original contributions not published elsewhere in whole or major part and include reports of research as well as reviews since symposia may embrace both types of presentation.

CONTENTS

PREFACE

As computing hardware and software continues to pervade the various areas of chemical research, education, and technology, various important developments begin to emerge. For example, for areas in which large "number crunching" is required, larger and faster computing systems have been developed that incorporate parallel processing, which have provided substantial increases in speed of problem solving compared with sequential processing. In other areas, such as data acquisition and equipment control, minicomputers and "midicomputers" have been designed and built to provide substantial improvements in both the quality of the data collected and the implementation of new experiments that could not be performed without the computer system assistance. Equally important developments in software have also evolved, from the implementation of convenient timesharing systems for program development to the development of a variety of application program "packages" for use in various chemical research areas.

While the limits achievable through better hardware design or more efficient programming of available algorithms are far from being reached, it is now becoming apparent that the algorithms themselves may present both substantial difficulties and opportunities for significant progress. In other words, it may no longer be a feasible strategy to assume that either a faster computer or a more efficiently programmed existing algorithm will be adequate in solving a given problem.

To focus more clearly on this emerging area of importance, a symposium was organized as a part of the Fall American Chemical Society Meeting in San Francisco, on August 30, 1976. The goal was to bring together several experts in the development of algorithms for chemical research so that the state of the art might be assessed. These persons, whose papers are included in this volume, discussed not only the significant developments in algorithms that have already occurred, but also indicated places where currently available algorithms were not adequate.

While it is not possible in a single symposium to discuss the entire spectrum of areas where significant algorithmic development has occurred or is needed, an attempt was made to include several of the important areas where progress is evident. In particular, the papers in this volume include discussions of the use of graph theory in algorithm design, algorithm design and choice in quantum chemistry, molecular scattering, solid state description and pattern recognition, and the handling of chemical information. As both the authors and the topics indicate, the general topic is extremely diverse in scope, involving expertise from several disciplines in the search for new and improved algorithms. While this area is currently in its infancy, its potential impact is great, and it is hoped that these papers will serve both as a reference to the current state of the art and as an impetus to extend the study of algorithmic development to other areas as well.

The University of Kansas   RALPH E. CHRISTOFFERSEN
Lawrence, Kansas
December 1976

Graph Algorithms in Chemical Computation

ROBERT ENDRE TARJAN*

Computer Science Dept., Stanford University, Stanford, CA 94305

1.   Introduction.

The use of computers in science is widespread. Without powerful number-crunching facilities at his** disposal, the modern scientist would be greatly handicapped, unable to perform the thousands or millions of calculations required to analyze his data or explore the implications of his favorite theory. He (or his assistant) thus requires at least some familiarity with computers, the programming of computers, and the methods which might be used by computers to solve his problems. An entire branch of mathematics, numerical analysis, exists to analyze the behavior of numerical algorithms.

However, the typical scientist's appreciation of the computer may be too narrow. Computers are much more than fast adders and multipliers; they are symbol manipulators of a very general kind. A scientist who writes programs in FORTRAN or some similar, scientifically oriented computer language, may be unaware of the potential use of computers to solve computational, but not necessarily numeric, problems which might arise in his research.

This paper discusses the use of computers to solve non-numeric problems in chemistry. I shall focus on a particular problem, that of identifying chemical structure, and examine computer methods for solving it. The discussion will include

---

*   This research was partially supported by the National Science Foundation, grant MCS75-22870, and by the Office of Naval Research, contract N00014-76-C-0688.

**   For the purpose of smooth reading, I have used the masculine gender throughout this paper.

elements of graph theory, list processing, analysis of algorithms, and computational complexity. I write as a computer scientist, not as a chemist; I shall neglect details of chemistry in order to focus on issues of algorithmic applicability, simplicity, and speed. It is my hope that some readers of this paper will become interested in applying to their own problems in chemistry the methods developed in recent years by computer scientists and mathematicians.

The paper is divided into several sections. Section 2 discusses representation of chemical molecules as graphs. Section 3 covers complexity measures for computer algorithms. Section 4 surveys what is known about the structure identification problem in general. Section 5 solves the problem for molecules without rings. Section 6 gives a method for analyzing a molecule by systematically breaking it into smaller parts. Section 7 discusses the case of "planar" molecules. Section 8 outlines a complete method for structure identification, and mentions some further applications of the ideas contained herein to chemistry.

2.   Molecules and Their Representation.

Consider a hypothetical chemical information system which performs the following tasks. If a chemist asks the system about a certain molecule, the system will respond with the information it has concerning that molecule. If the chemist asks for a listing of all molecules which satisfy certain properties (such as containing certain radicals), the system will respond with all such molecules known to it. If the chemist asks for a listing of possible molecules (known or not), which satisfy certain properties, the system will provide a list.

Such an information system must be able to identify molecules on the basis of their structure. Given a molecule, the system must derive a unique code for the molecule, so that the code can be looked up in a table and the properties of the molecule located. It is this coding or cataloging problem which I want to consider here. A number of codes for molecules have been proposed and used; e.g. see (1,2,3,4). The existence of many different codes with no single standard suggests the importance and the difficulty of the problem. I shall attempt to explain why the problem is difficult, and to suggest some computer approaches to it.

To deal with the problem in a rigorous fashion, we couch it within the branch of mathematics called graph theory. A graph $G = (V, E)$ is a finite collection V of vertices and a finite collection E of edges. Each edge (v, w) consists of an unordered pair of distinct vertices. Each edge and each vertex may in addition have a label specifying certain information

about it. We represent a chemical molecule as a graph by
constructing one vertex for each atom and one edge for each
chemical bond; a ball-and-stick model of a molecule is really a
graph representation of it. We label each vertex with the type of
atom it represents. See Figure 1 for an example.

Two vertices v and w of a graph are said to be <u>adjacent</u>
if (v,w) is an edge of the graph. If (v,w) is an edge, and
v is a vertex contained in it, the edge and vertex are said to
be <u>incident</u>. Two graphs $G_1 = (V_1, E_1)$ and $G_2 = (V_2, E_2)$ are
said to be <u>isomorphic</u> if their vertices can be identified in a
one-to-one fashion so that, if v_1 and w_1 are vertices in G_1
and v_2 and w_2 are the corresponding vertices in G_2, then
(v_1, w_1) is an edge of G_1 if and only if (v_2, w_2) is an edge
of G_2. Furthermore the pairs v_1, v_2; w_1, w_2; and
(v_1, w_1), (v_2, w_2) must have the same labels if the graphs are
labelled.

The problem we shall consider is this: given two graphs,
determine if they are isomorphic. Or: given a graph, construct
a code for it such that two graphs have the same code if and only
if they are isomorphic. Notice that this mathematical abstraction
of chemical structure identification neglects some details of
chemistry. For instance, we allow bonds between only two molecules, thereby precluding the representation of resonance structures, and we ignore issues of stereochemistry (if two bonds of a
carbon atom are fixed, our model allows free interchanging of the
other two, whereas in the real world such interchanging may
produce stereoisomers; see Figure 2). However, these are
differences of detail only, which can easily be incorporated into
the model; we neglect them only for simplicity. Note also that
our model does not allow loops (edges of the form (v,v)), but
it <u>does</u> allow multiple edges (which may be used to represent
<u>multiple</u> bonds, or for other purposes).

A generalization of the isomorphism problem is the <u>subgraph</u>
<u>isomorphism</u> problem. Given two graphs $G_1 = (V_1, E_1)$ and
$G_2 = (V_2, E_2)$, we say G_1 is a subgraph of G_2 if V_1 is a
subset of V_2 and E_1 is a subset of E_2. The subgraph
isomorphism problem is that of determining if a given graph G_1
is isomorphic to a subgraph of another given graph G_2. This is
one of the problems our hypothetical information system must solve
to provide a list of molecules containing certain radicals. We
shall deal with this problem briefly; it seems to be much harder
than the isomorphism problem.

If a computer is to efficiently encode molecules it must
first have a way to represent a molecule, or a graph. We consider

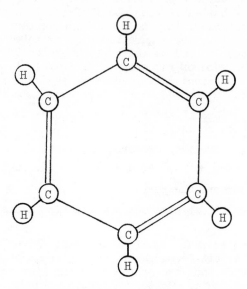

Figure 1.   Graphic representation of benzene

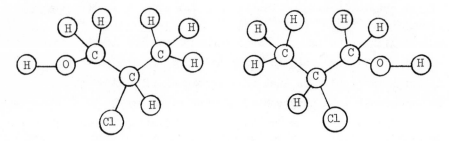

Figure 2.   Stereoisomers

two standard ways to represent graphs in a computer. The first is
by an adjacency matrix. If $G = (V, E)$ is a graph with n
vertices numbered from 1 to n , an adjacency matrix for G is
the n by n matrix $M = (m_{ij})$ with elements 0 and 1 , such
that $m_{ij} = 1$ if (v_i, v_j) is an edge of G and $m_{ij} = 0$ otherwise. See Figure 3(a), (b). Note that M is symmetric and that
its main diagonal is zero. The matrix M is not a code for G
since it is not unique; it depends upon the vertex numbering.

An adjacency matrix representation of a graph has several
nice properties. Many natural graph operations correspond to
standard matrix operations (see (5) for some examples). The bits
of M can be packed in groups into computer words, so that
storage of M requires only n^2/w words, if w is the word
length of the machine (or only $n^2/2w$ words, if advantage is
taken of the symmetry of M). If M is packed into words in
this way, the bits can be processed w at a time, at least in
certain kinds of computations.

However, the matrix representation has some serious disadvan-
tages. An important property of graphs representing chemical
molecules is that they are sparse; most of the potential edges are
missing. Since each atom has a fixed, small valence, the number of
edges in a graph representing a molecule is no more than
some fixed constant times n , the number of vertices. However,
in an arbitrary graph the number of edges can be as large as
$(n^2-n)/2$ (or larger, if there are multiple edges). An adjacency
matrix for a sparse graph contains mostly zeros, but there is no
good way of exploiting this fact. It has been proved that testing
many graph properties, including isomorphism, requires examining
some fixed fraction of the elements of the adjacency matrix in the
worst case (6). Any algorithm which uses a matrix representation
of a graph thus runs in time proportional to at least n^2 in the
worst case. If we wish to deal with large graphs and hope to get
a running time close to linear in the size of the graph, we must
use a different representation.

The one we choose is an adjacency structure. An adjacency
structure for a graph $G = (V, E)$ is a set of lists, one for each
vertex. The list for vertex v contains all vertices adjacent
to v . Note that a given edge (v, w) is represented twice;
w appears in the adjacency list for v and v appears in the
adjacency list for w . See Figure 3(c).

An adjacency structure is surprisingly easy to define and
manipulate in FORTRAN or any other standard programming language.
We use three arrays, which we may call adjacent to, vertex, and
next. For any vertex v , the element $e_1 = $ adjacent to (v)
represents the first element on the adjacency list for vertex v .
The corresponding vertex is vertex(e_1) , and the element

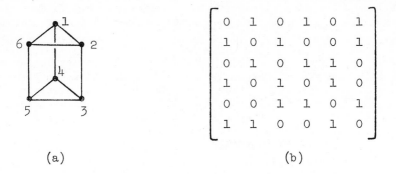

(a)   (b)

1:   2, 4, 6
2:   1, 3, 6
3:   2, 4, 5
4:   1, 3, 5
5:   3, 4, 6
6:   1, 2, 5

(c)

	1	2	3	4	5	6
adjacent to:	1	2	8	4	14	6

	1	2	3	4	5	6	7	8	9	10	11	12	13	14	15	16	17	18
vertex:	2	1	4	1	6	1	3	2	6	2	4	3	5	3	5	4	6	5
next:	3	7	5	12	/	10	9	11	/	18	13	15	/	16	/	17	/	/

(d)

Figure 3. Graphic representations: (a) graph, (b) adjacency matrix, (c) adjacency structure, and (d) array representation of adjacency structure

$e_2 = \underline{next}(e_1)$ represents the next element on the list. A null
element indicates the end of the list. See Figure 3(d). The
total amount of storage required by these arrays is $n+4m$, where
n is the number of vertices in the graph and m is the number of
edges; the total storage is thus linear in the size of the graph.
Searches and other natural graph operations are easy to implement
using such a data structure; e.g. see (7, 8). If the graph is
labelled we can use two extra arrays which give vertex and edge
labels. Athough the matrix representation of a graph is simple
and mathematically elegant, the adjacency structure representation
seems to be much more useful for computers.

3.   Notions of Complexity.

If we are to discuss computer methods, we need some way of
measuring the performance of an algorithm. We would like our
code for molecules to be simple, natural, and easy to compute.
Concepts like "simple" and "natural", although very important in
any real-world cataloguing system, are difficult to define and
quantify. We shall use a measure based on a machine's point of
view, rather than on a human's. Though an algorithm good by such
a measure may be unwieldy for human use, at best a method useful
for machines will also be useful for people. At worst, such a
measure provides a firm base for discussion of the merits of
various methods.
One possible measure of algorithmic complexity is program
size. Such a measure is related to the inherent simplicity or
complexity of a method. This measure is static; it is independent
of the size or structure of the particular input data. Some other
possible measures are dynamic; they measure the amount of a
resource used by the method as a function of the size of the input
data. Typical dynamic measures are running time and storage
space.
Program size as a measure has the disadvantage that in many
cases the simplest algorithm is a brute force examination of all
possibilities; the running time of such an algorithm is exponential in the size of the input and thus only very small graphs can
be analyzed. The algorithms we shall consider all use storage
space linear or quadratic in the number of vertices in the input
graph; thus storage space as a measure does not discriminate
finely enough for our purposes. The running time of an algorithm
is strongly related to the algorithm's usefulness if it is run
many times. We therefore choose running time as a function of
input size as our measure of complexity.
How shall we measure running time? One possibility is to run
the program several times on various sets of input data and
extrapolate. This approach is very dangerous. If the number of
examples tried is too small, the extrapolation is probably
meaningless. If the number of examples tried is large and drawn

from a suitably defined random population, the extrapolation may
be statistically meaningful. However, defining a random graph
in a way which is realistic for chemistry is a very tricky
problem. Furthermore any statistical method may miss rare but
very bad cases; we would not like our cataloguing system to spend
hours on an occasional bizarre molecule. We are therefore only
satisfied with a careful theoretical analysis of an algorithm
leading to a worst-case bound on its running time.

To account for variability in machines, we ignore constant
factors and pay attention only to the asymptotic growth rate of
the running time as a function of the size of the problem graph.
Our measure is thus machine independent and most valid for large
graphs. If machine-dependent constant factors and running time
on small graphs are of interest, computer experiments or a more
detailed analysis must be used. For convenience, we shall use the
notation " $f(n)$ is $O(g(n))$ " to denote that the function $f(n)$
satisfies $f(n) \leq cg(n)$ for some positive constant c and all
n , where f and g are non-negative functions of n .

4.   Isomorphism and Subgraph Isomorphism.

The isomorphism problem for general graphs is not an easy
one. Given two graphs G_1 and G_2 of n vertices, the number
of possible one-to-one mappings of vertices is $n!$, and a brute
force approach, which tries all the possibilities, is too time-
consuming except for small graphs. A backtracking search (9),
fares somewhat better. Initially, one vertex from each graph is
chosen, and these vertices are matched. In general, some vertex
w_1 adjacent to an already-matched vertex v_1 in G_1 is chosen
and matched with some vertex w_2 adjacent to the vertex v_2 in
G_2 previously matched to v_1 . Then w_1 and w_2 are compared
to make sure their adjacencies with already-matched vertices are
consistent. If so, a new vertex for matching is chosen. If not,
the last matched pair is unmatched and a new matching tried.
The process continues until either all vertices are matched or
there is found to be no way of matching the vertex sets of the
two graphs.

Backtrack search saves time over the brute force method by
abandoning an attempt at matching as soon as it is known to fail.
The running time of backtrack search depends in a complicated way
upon the structure of the graph; the best we can say in general is
that if d is the maximum valence (number of vertices adjacent to
a given vertex) in either graph, the maximum running time of backtrack search is $O((d-1)^n)$ -- still exponential, but better than
brute force.

The most successful algorithms for general graph isomorphism
use the backtrack approach (as a fall-back method) in combination

with a partitioning method (10,11,12,13). The idea is to partition
the combined vertex sets of the two graphs so that any isomorphic
mapping between the graphs preserves the partitioning. The method
has four main steps.

1. Choose an initial partition of the vertex sets.
2. Refine the partition. If any subset of the partition
    contains more vertices from one graph than from the other,
    go to step 4.
3. If each subset of the partition contains a single vertex
    from each graph, try the implied matching to see if it gives
    an isomorphism. If it does, halt with the isomorphism; if
    not, go to step 4. If some subset contains two or more
    vertices from one graph, choose a vertex in this subset from
    each graph, match these vertices, and go to step 2 (the new
    matching allows further refinement of the partition).
4. Backtrack.   Back up to the partition existing when the
    last match was made. Try a new match and go to step 2. If
    all matches have been tried, back up to the previous match.
    If all possibilities for the very first match have been
    tried, halt. The graphs are not isomorphic.

For the initial partition we divide vertices up according to
their labels and their valences. Other more elaborate
partitionings are possible; see (14,15).

We carry out the refinement step in the following way. For
each vertex, we determine the number of adjacent vertices in each
subset of the partition. This information itself partitions the
vertices. We take the intersection of this partition with the old
partition as our new partition. We repeat this refining step
until no further refinement takes place. Implementation of the
repeated refinement step is somewhat tricky; Hopcroft (16) has
provided a good implementation. The effect of matching two
vertices in step 3 is to place them by themselves in a new subset
of the partition. Thus step 3 guarantees refinement of the
partition. See Figure 4 for an example of the application of the
algorithm.

The idea behind this algorithm is to use all possible local
means of distinguishing between vertices before guessing a match.
The method seems to work quite well in practice. It is possible
that some version of this partitioning method has a time bound
which is a polynomial function of n. (To prove this requires
showing that the amount of backtracking is polynomial in n; the
refinement step requires only $O(m \log m)$ time, where m is the
number of edges, if Hopcroft's implementation is used.) However,
the present theoretical bounds on the algorithm are no better than
those for backtrack search. It is a major open question whether
a polynomial-time algorithm exists for the general graph
isomorphism problem.

The situation for the subgraph isomorphism problem is somewhat better understood and somewhat more gloomy. It is possible

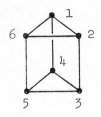 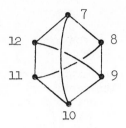

(a)

(b)   {1,2,3,4,5,6,7,8,9,10,11,12}
              A: valence 3

(c)   {1,7} {2,3,4,5,6,8,9,10,11,12}
         B   C

(d)   {1,7} {2,4,6,8,10,12}{3,5,9,11}
         B   D: 1B, 2C   E: 3C

(e)   {1,7}   {2,6}   {4,8,10,12}   {3,5}   {9,11}
         B   F: 1B, 1D, 1E   G: 1B, 2E   H: 2D, 1E   I: 1B, 2D

Figure 4. Isomorphism test by partitioning: (a) graphs, (b) initial partition,(c) initial match 1–7, (d) first refinement, and (e) further refinement (match fails since F contains no vertices of second graph). Complete test requires matching 1 successively to 8, 9, 10, 11, 12, failing each time.

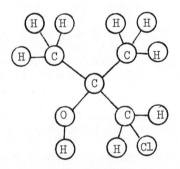

Figure 5. A tree

to generalize the partitioning algorithm described above so that
it solves the subgraph isomorphism problem (17). However, the
results of this method in practice seem to be mixed. Furthermore
it has been proved that the subgraph isomorphism problem belongs
to a class of problems called NP-complete. The NP-complete
problems include a number of well-studied, apparently hard
problems such as the travelling salesman problem of operations
research, the tautology problem of propositional calculus, and
many other combinatorial problems. The NP-complete problems have
the property that if any one of them has a polynomial-time
algorithm, they all do. Since no one has discovered a polynomial-
time algorithm for any of these problems, though many people have
tried, it seems likely that none of these problems is solvable in
polynomial time. It is not known whether the graph isomorphism
problem itself is NP-complete. For a discussion of NP-complete
problems, see (18,19,20).

It would seem that our attempt to solve the graph isomorphism
problem with a provably good algorithm is doomed to failure, and
that we must be satisfied with a heuristic; that is, with a method
which seems to work well in many cases for reasons which we do not
understand. However, by lowering our sights somewhat, we can go a
long way toward a solution which is both practical and theoretically efficient. We shall first consider the isomorphism problem
for trees. For such graphs, there is a good isomorphism
algorithm. Next, we study a decomposition method for representing
a graph as a collection of smaller graphs joined in a tree-like
fashion. We then examine the important special case of planar
graphs. Finally, we combine these ideas to produce an isomorphism
algorithm which is very fast on planar graphs and is likely to
work well on most, if not all, chemical molecules.

5. Codes for Trees.

Let $G = (V, E)$ be a directed graph. A simple path from a
vertex v_1 to a vertex v_k in G is a sequence of distinct
edges (v_1, v_2), (v_2, v_3), ..., (v_{k-1}, v_k). The length of the path
is $k-1$, the number of edges it contains. A cycle is a simple
path from a vertex v_1 to itself. A graph is connected if every
pair of vertices is joined by a path. In the description of a
backtrack search in Section 4 we implicitly assumed that the
graphs of interest were connected; we shall continue to make this
assumption. A tree is a connected graph with no cycles (see
Figure 5 for an example).

In contrast to the isomorphism problem for general graphs,
the isomorphism problem for trees is relatively easy. Any tree
with n vertices has exactly $n-1$ edges. We shall describe an
algorithm for constructing, in $O(n)$ time, a code for any tree,
such that two trees are isomorphic if and only if they have

identical codes. Variants of the algorithm have appeared in many
places (21,22,23,24) and it has in fact been used in chemical
computation (25).

To extract a unique code for a tree we must first put the
tree into a canonical form. The first step in doing this is to
find a uniquely determined vertex or edge in the tree. A tree
has at least two vertices of valence one. We call such vertices
leaves. For a given vertex v, let the height $h(v)$ of v be
the length of the longest path from v to a leaf. A tree
contains either a unique vertex of largest height, or two
adjacent vertices of largest height (26). Since height must be
preserved under isomorphism, this unique vertex or pair of
vertices can be used as a starting point for construction of the
canonical tree. If there are two vertices of largest height, we
add a new vertex in the middle of the edge joining them and label
it as a dummy vertex. Then we can assume our tree always has a
unique vertex of largest height, which we call the root.

Each vertex v except the root has a unique parent u which
is adjacent to v and satisfies $h(u) \geq h(v)+1$. All other
vertices w adjacent to v are called its children and satisfy
$h(w) \leq h(v)-1$. We define ancestors and descendants in the
obvious way. Each vertex v in the tree defines a subtree
consisting of v and its descendants (see Figure 6).

We define a total ordering with vertex labels by the
following rules.

(1) If T and U are two trees with different labels on their
     roots, order the trees according to the labels of the roots.

(2) If T and U are two trees with the same label on their
     roots, let $T_1, T_2, ..., T_k$ be the subtrees defined by the

     children of the root of T (in increasing order) and let
     $U_1, U_2, ..., U_\ell$ be the subtrees defined by the children of the

     root of U. If there is some index j such that T_i is

     isomorphic to U_j for $i < j$ and T_j is less than U_j,

     or if T_i is isomorphic to U_i for $1 \leq i \leq k$ and $k < \ell$,

     then T is defined to be less than U.

That is, to compare two trees, we first compare their root
labels. If these are identical, we order the subtrees defined by
the children of the roots, and compare the ordered sequences of
subtrees lexicographically.

Using this ordering, we can construct a canonical representation of a given tree by reordering the children of each vertex
according to the order defined above. See Figure 6. From this
canonical representation, we can construct a linear code which
represents the tree uniquely. There are many possible ways to do
this; one way is defined by the following rules.

(1)   The code $\underline{\text{code}}$(T) for a tree T consisting of a single
      vertex is $\underline{\text{its}}$ label.

(2)   If T is a tree of more than one vertex, and $T_1, T_2, \ldots, T_k$
      are the subtrees defined by the children of the roots of T
      (in order), then the code for T is
      $$\underline{\text{code}}(T) = \underline{\text{code}}(\underline{\text{root}})(\underline{\text{code}}(T_1)\underline{\text{code}}(T_2) \ldots \underline{\text{code}}(T_k)) .$$
      For instance, the code for the molecule in Figure 6 is
      C(C(ClHH)C(HHH)C(HHH)O(H)).

This method gives a unique code for each tree; two trees are
isomorphic if and only if they have the same code (we have
neglected to include edge labels in the code, but it is easy to do
so if necessary). The code is quite natural, and it is easy to
reconstruct a tree given its code. The reordering of subtrees is
what guarantees that each tree has only one code. One can vary
the exact definition of the ordering; what is important is that
the subtrees be ordered $\underline{\text{somehow}}$. When this algorithm is applied
to chemical molecules, it is useful to use abbreviations in the
code, such as omitting explicit reference to hydrogen atoms; e.g.
see (27).

Implementing the reordering algorithm is somewhat
complicated, since the sorting requires comparison of sequences
element-by-element. See (28) for a good implementation.
Constructing the code for a tree of n vertices requires O(n)
time with this implementation. We can expect to find no faster
algorithm, since any method must inspect the entire tree.

On trees, not only is the isomorphism problem efficiently
solvable, but so is the subgraph isomorphism problem. Edmonds
and Matula (29) have discovered an algorithm which will determine
whether one tree is isomorphic to a subtree of another in
$O(n^{5/2})$ time, where n is the number of vertices in the larger
tree. This bound can be improved substantially if the valence
of all vertices is bounded by a small constant. The algorithm
may be of practical value, but this has yet to be tested.

6.   Decomposition by Connectivity.

Though the algorithm of Section 5 for encoding trees is
simple and fast, most chemical molecules are not trees. However,
they $\underline{\text{are}}$ quite sparse and often tree-like. Our approach in this
section will be to represent an arbitrary graph as a number of
pieces linked in tree-like fashion. We can then encode the graph
by encoding each piece separately, using these codes as labels on
the linkage tree, and applying the tree encoding algorithm of
Section 5 to encode the entire graph. In this way we can make the
most out of our tree encoding method; the non-tree-like parts of
the graph will usually be small.

To decompose a graph, we determine its $\underline{\text{connectivity}}$. Let
G = (V, E) be a connected graph. A $\underline{\text{cut set}}$ of G is a subset

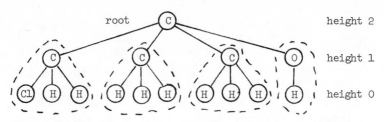

Figure 6. *Tree of Figure 5 in canonical form. Dashes enclose subtrees of children of the root.*

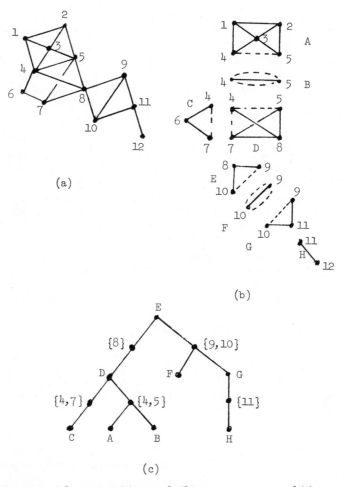

(a)

(b)

(c)

Figure 7. *Schematic of (a) a graph, (b) its components, and (c) its decomposition tree*

of vertices S such that there are at least two vertices v
and w (not in S) for which every path from v to w passes
through a vertex in S . Removal of the vertices in S thus
breaks G into two or more connected pieces. If we add the
vertices in G to each piece, the resultant subgraphs of G
are called the components of G with respect to the cutset S .
    We concentrate on cutsets containing no more than two
vertices. By applying the following procedure, we break G into
a number of smaller graphs.

    Decomposition algorithm.   Begin with a single component
consisting of the entire graph. Repeat the following step until
it no longer applies:
    Find a cutset of size one or two in some component. If it is
a cutset of size one, subdivide the component into its components
with respect to the cutset. If it is a cutset of size two, say
{v,w} , subdivide the component into its components with respect
to the cutset, and add a new (dummy) edge (v,w) to each new
component.

    The importance for isomorphism testing of this algorithm is
three-fold: first, the components found by the algorithm are
essentially unique (preserved under isomorphism). (To guarantee
uniqueness we must slightly modify the definition of components
with respect to cutsets of size two; see (30,31,32).   Second, the
way the components fit together can be represented by a decomposition tree (33). This tree contains one vertex for each
component and one vertex for each cutset. A cutset is adjacent
to a component in the tree if the vertices of the cutset are in
the component. Figure 7 gives an example of a graph, its
components, and its decomposition tree.
    Third, it is easy to find the components and the decomposi-
tion tree. An algorithm for this purpose, which uses depth first
search (a systematic method of exploring a graph) has been
developed (34,35,36). It runs in O(n+m) time on an n vertex,
m edge graph.
    Each component with respect to the decomposition is of one
of three kinds -- a bond (single edge or set of multiple edges),
a cycle, or a graph with no multiple edges and no cutsets of
size one or two, called a triconnected graph. It is easy to
encode bonds and cycles; all that is missing is a method of
encoding triconnected graphs. If we can encode all the
components, we can use the resultant codes as labels in the
decomposition tree and apply the Section 5 algorithm to encode
the entire tree. The running time of this algorithm will be
O(n+m) for everything except the encoding of the triconnected
components. If we use the partitioning method of Section 4 as a
basis for encoding triconnected components, the complete algorithm
will probably do quite well in practice. However, we have one
more improvement to consider.

7.   Planar Graphs.

A planar graph is a graph which can be drawn on a piece of
paper in such a way that no edges cross. Most chemical molecules
(with the possible exception of complex organic molecules) are
planar (note that this does not mean planar in the sense of
stereochemistry). For planar graphs the isomorphism problem also
has an easy solution.

When a graph is drawn in the plane, the drawing specifies a
circular ordering of the edges around each vertex. A triconnected
graph has the property that, if it is planar, its planar representation is unique up to mirror image. Thus there are only two ways
of drawing a triconnected planar graph in the plane (two ways of
specifying the circular ordering of edges around each vertex).

We can use this uniqueness to derive a code for any planar
triconnected graph. First, we represent the graph in the plane.
This can be done in $O(n)$ time (37). Next, we encode it. One
way to do this was suggested by Weinberg (38). We explore the
graph in the following way. We pick some starting edge and
traverse it from one end to the other. When reaching the other
end, we choose the next edge clockwise around the vertex and
traverse it. We continue traversing edges in this way. Whenever
we reach a vertex reached previously, we back up along the most
recently traversed edge and pick the next edge clockwise. We
continue the search until we have traversed each edge in both
directions and returned to our starting point.

Such a search is uniquely determined by the choice of the
starting edge and the direction to traverse it. We can construct a
linear code during the search by writing a number (and a label)
for each vertex reached, numbering the first vertex one, the next
two, and so on. See Figure 8. To get a unique code, we construct
a code for each possible edge and direction of traversal, for each
of the two planar representations of the graph. Then we choose
the lexicographically smallest of all the possible codes. A triconnected planar graph of $n \geq 3$ vertices has at most $3n-6$
edges (39), so we generate at most $12n-24$ codes, each of length
n , and the total time to get a unique code is $O(n^2)$.

This encoding algorithm is very easy to program, but it is
possible to get a faster algorithm by using more sophisticated
methods. Hopcroft's partitioning algorithm (40) can be used to
encode triconnected planar graphs in $O(n \log n)$ time (41).
Hopcroft and Wong (42) have devised a very complicated algorithm
which will encode a triconnected planar graph in $O(n)$ time.
More recently, Fontet (43) has devised a simpler $O(n)$ -time
encoding algorithm. The practicality of these algorithms has
yet to be tested.

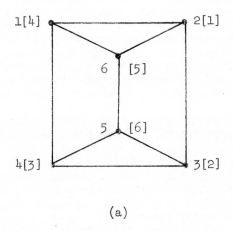

(a)

(b) 1 23 41 45 61 62 65 35 43 21

(c) 1 23 41 45 15 62 63 65 43 21

Figure 8. (a) Planar Graph. (b) Code extracted by search starting with edge (1, 2). (Vertices are numbered in search order.) (c) Code extracted by search starting with edge (2, 3). (Numbers in brackets give the numbering for this search.) Code (c) is chosen since it is smaller lexicographically. All other codes are identical to either (b) or (c).

8.   Summary and Other Applications.

   We are now in a position to outline a complete isomorphism
algorithm. We test isomorphism of two graphs by encoding each
graph and testing the codes for equality. To encode a graph, we
decompose it by finding all cutsets of size one and two, and
forming the corresponding components and decomposition tree. We
encode each bond component and each cycle component in some
obvious way. We encode each triconnected component as follows.
We test the component for planarity. If it is planar, we encode
it using one of the methods in Section 7. If it is not planar,
we encode it using the partitioning algorithm of Section 4. We
use the codes for components as labels in the decomposition tree,
and encode the tree (and thus the entire graph) using the method
of Section 5.
   The overall result is a method with a running time of $O(n+m)$
on n-vertex, m-edge graphs, plus whatever time is required to
encode non-planar triconnected components. Though this algorithm
has many parts, and programming it is quite a job, it has the
potential to be of practical value. Though most of the parts of
the algorithm have been programmed individually, the complete
algorithm has not been programmed. Hopefully, this situation will
be remedied in the near future.
   Though the isomorphism problem is a formidable one, we have
examined some ideas and some methods which can go a long way
toward solving it. Many of the ideas we have considered have
applications in other areas of chemistry. For instance, we have
discussed representing a sparse graph as an adjacency matrix with
many zeros. We can turn this idea around and use a graph to
represent a sparse matrix (the matrix elements become labels for
the corresponding graph edges). We can then apply graph-
theoretic techniques to matrix problems such as solving a system
of linear equations and computing eigenvalues and
A large literature has developed in this area; see (44,45,46) for
instance.
   Another application of graph theory to chemistry is in
chromosome analysis. Suppose a chromosome is broken into a
number of pieces and each piece analyzed. If this is done a
number of times, the pieces found will overlap in various ways.
The problem is to use the overlap information to reconstruct the
entire chromosome. For linear chromosomes, a linear-time
algorithm has been developed to solve this problem (47,48). For
chromosomes which are rings, the problem seems surprisingly to
be much harder and no good algorithm is known (49).

(1) "Survey of Chemical Notation Systems," National Academy of
     Sciences, National Research Council Publication 1150, 1964.
(2) Lederberg, J., "Dendral-64, a System for Computer Construc-
     tion, Enumeration, and Notation of Organic Molecules as Tree
     Structures and Cyclic Graphs, Part I," NASA Scientific and
     Technical Aerospace Report, STAR No. N65-13158 and CR 57029,
     1964.
(3) Lederberg, J., "Dendral-64, a System for Computer Construc-
     tion, Enumeration and Notation of Organic Molecules as Tree
     Structures and Cyclic Graphs, Part II," NASA Scientific and
     Technical Aerospace Report, STAR No. N66-14074 and CR 68898,
     1965.
(4) Sussenguth, E., Jr., J. Chem. Doc. (1965) $\underline{5}$, 36-43.
(5) Harary, F., "Graph Theory," 150-151, Addison-Wesley,
     Reading, Mass., 1969.
(6) Rivest, R. and Vuillemin, J., Seventh ACM Symp. on Theory of
     Computing (1975), 6-11.
(7) Hopcroft, J.E. and Tarjan, R. E., Comm. ACM (1973) $\underline{16}$,
     372-378.
(8) Tarjan, R., SIAM J. Comput. (1972) $\underline{1}$, 146-160.
(9) Berztiss, A. T., Journal ACM (1973) $\underline{20}$, 365-377.
(10) Corneil, D. G. and Gotlieb, C. C., Journal ACM (1970) $\underline{17}$,
     51-64.
(11) Schmidt, D. C. and Druffel, L. E., Journal ACM (1976) $\underline{23}$,
     433-445.
(12) Sussenguth, E., Jr., J. Chem. Doc. (1965) $\underline{5}$, 36-43.
(13) Unger, S. H., Comm. ACM (1964) $\underline{7}$, 26-34.
(14) Corneil, D. G. and Gotlieb, C. C., Journal ACM (1970) $\underline{17}$,
     51-64.
(15) Schmidt, D. C. and Druffel, L. E. Journal ACM (1976) $\underline{23}$,
     433-445.
(16) Hopcroft, J. E., in Kohavi, Z. and Paz, A., eds., "Theory of
     Machines and Computations," 189-196, Academic Press,
     New York, 1971.
(17) Sussenguth, E., Jr., J. Chem. Doc. (1965) $\underline{5}$, 36-43.
(18) Cook, S., Third ACM Symp. on Theory of Computing (1971),
     151-158.
(19) Karp, R. M., in Miller, R. E. and Thatcher, J. W., eds.,
     "Complexity of Computer Computations," 85-104, Plenum Press,
     New York, 1972.
(20) Karp, R. M., Networks (1975) $\underline{5}$, 45-68.
(21) Busacker, R. G. and Saaty, T. L., "Finite Graphs and
     Networks: An Introduction with Applications," 196-199,
     McGraw-Hill, New York, 1965.
(22) Lederberg, J., NASA Scientific and Technical Aerospace
     Report, STAR No. N65-13158 and CR 57029, 1964.
(23) Scoins, H. I., Machine Intelligence (1968) $\underline{3}$, 43-60.
(24) Weinberg, L., Proc. Third Annual Allerton Conf. on Circuit
     and System Theory (1965), 733-744.

(25) Lederberg, J., NASA Scientific and Technical Aerospace
      Report, STAR No. N65-13158 and CR 57029, 1964.
(26) Harary, F., "Graph Theory," 35-36, Addison-Wesley, Reading,
      Mass., 1969.
(27) Lederberg, J., NASA Scientific and Technical Aerospace
      Report, STAR No. N65-13158 and CR 57029, 1964.
(28) Aho, A. V., Hopcroft, J. E., and Ullman, J. D., "The Design
      and Analysis of Computer Algorithms," 84-86, Addison-Wesley,
      Reading, Mass., 1974.
(29) Matula, D. W., SIAM Review (1968) $\underline{10}$, 273-274.
(30) Hopcroft, J. E. and Tarjan, R. E., SIAM J. Comput. (1973) $\underline{2}$,
      135-158.
(31) Maclaine, S., Duke Math. J. (1937) $\underline{3}$, 460-472.
(32) Tutte, W. T., "Connectivity in Graphs," University of
      Toronto Press, Toronto, 1966.
(33) Harary, F., "Graph Theory," 36-37, Addison-Wesley, Reading,
      Mass., 1969.
(34) Hopcroft, J. E. and Tarjan, R. E., SIAM J. Comput. (1973) $\underline{2}$,
      135-158.
(35) Hopcroft, J. E. and Tarjan, R. E., Comm. ACM (1973) $\underline{16}$,
      372-378.
(36) Tarjan, R. E., SIAM J. Comput. (1972) $\underline{1}$, 146-160.
(37) Hopcroft, J. E. and Tarjan, R. E., Journal ACM (1974) $\underline{21}$,
      549-568.
(38) Weinberg, L., IEEE Trans. on Circuit Theory (1966) $\underline{CT-13}$,
      142-148.
(39) Harary, F., "Graph Theory," 104, Addison-Wesley, Reading,
      Mass., 1969.
(40) Hopcroft, J. E., in Kohavi, Z. and Paz, A., eds., "Theory of
      Machines and Computations," 189-196. Academic Press,
      New York, 1971.
(41) Hopcroft, J. E. and Tarjan, R. E., Journal of Computer and
      System Sciences (1973) $\underline{7}$, 323-331.
(42) Hopcroft, J. E. and Wong, J. K., Sixth Annual ACM Symp. on
      Theory of Computing (1973), 172-184.
(43) Fontet, M., Proc. Third International Colloquium on
      Automata, Languages, and Programming, to appear.
(44) Bunch, J. R. and Rose, D. J., eds., "Sparse Matrix
      Computations," Academic Press, New York, 1976.
(45) Duff, I. S., "A Survey of Sparse Matrix Research," Technical
      Report CSS 528, Computer Science and Systems Division, AERE
      Harwell, 1976.
(46) Rose, D. J. and Willoughby, R., eds., "Sparse Matrices and
      their Applications," Plenum Press, New York, 1972.
(47) Benzer, S., Proc. of the National Academy of Sciences (1959)
      $\underline{45}$, 1607-1620.
(48) Lueker, G. S. and Booth, K. S., Seventh ACM Symp. on Theory
      of Computing (1975), 255-265.
(49) Booth, K. S., "P-Q Trees," Ph.D. Thesis, Dept. of Electrical
      Engineering and Computer Sciences, University of California,
      Berkeley, 1975.

Algorithm Design in Computational Quantum Chemistry

ERNEST R. DAVIDSON

Chemistry Dept., University of Washington, Seattle, WA 98195

Quantum chemistry is a diverse discipline which uses many different methods to correlate a wide variety of phenomena. In the earliest period of the subject the Schrödinger equation was solved exactly for a few simple model situations. These model solutions were then used to interpret the spectra, kinetics, and thermodynamics of molecules and solids.

During this period, accurate solutions for the electronic structure of helium (1) and the hydrogen molecule (2) were obtained in order to verify that the Schrödinger equation was useful. Most of the effort, however, was devoted to developing a simple quantum model of electronic structure. Hartree (3) and others developed the self-consistent-field model for the structure of light atoms. For heavier atoms, the Thomas-Fermi model (4) based on total charge density rather than individual orbitals was used.

Models for the electronic structure of polynuclear systems were also developed. Except for metals, where a free electron model of the valence electrons was used, all methods were based on a description of the elec-tronic structure in terms of atomic orbitals. Direct numerical solutions of the Hartree-Fock equations were not feasible and the Thomas-Fermi density model gave ridiculous results. Instead, two different models were introduced. The valence bond formulation (5) followed closely the concepts of chemical bonds between atoms which predated quantum theory (and even the discovery of the electron). In this formulation certain reasonable "configurations" were constructed by drawing bonds between unpaired electrons on different atoms. A math-ematical function formed from a sum of products of atomic orbitals was used to represent each configura-tion. The energy and electronic structure was then

found by the linear variation method (also called "resonance" or "configuration interaction"). Because of its
almost one-to-one correspondance with earlier chemical
concepts the valence bond model gained widespread accep-
tance (6). The molecular orbital model (7) assumed,
instead, that the electrons were in certain molecular
orbitals which could be expressed as linear combinations
of atomic orbitals. Configurations were then constructed as various ways of arranging electrons in orbitals. The molecular orbital model gave a clear interpretation of molecular spectra but was less transparent
than the valence bond method in modeling geometrical
structure of molecules (6,8). In almost all early applications of valence bond (9) and molecular orbital
(10) models the integrals encountered were too difficult
to actually evaluate so empirical values of the integrals were assumed which reproduced the phenomena being
studied.

   With the advent of the stored-program digital computer a minor revolution occurred in quantum chemistry.
The integrals appearing in the models being used for
small molecules were actually evaluated and it became
clear that molecules were enormously more complicated
than had been anticipated. The oversimplified valence
bond and molecular orbital methods often gave qualita-
tively ridiculous results when taken literally (11).

   As a consequence of these negative results, the
field of ab initio quantum chemistry developed with the
goal of finding computer algorithms for solving the
Schrödinger equation. The prospect of obtaining reliable results for molecular systems not susceptible to
direct measurement (repulsive potential energy surfaces,
upper atmosphere free radicals, etc.) and clarifying
the interpretation of experimental results which do not
follow simple models attracted interest in this field
in spite of the extraordinary expense of the approach
and the lack of chemical insight in the early results.

   In the ab initio approach the desired answers are
the experimental observables - spectral line positions,
shapes, intensities; scattering and reaction rates;
polarizabilities and optical rotary power; etc. These
are to be obtained from the Schrödinger equation by
numerical methods which are mathematically well-defined
and involve no intermediate parameters not appearing in
the Schrödinger equation itself.

   Usually the Born-Oppenheimer separation of nuclear
and electronic coordinates is assumed and small terms in
the hamiltonian, such as spin-orbit coupling, are
neglected in the first approximation. Perturbation

theory may be used to correct for these approximations by coupling electronic states in the next level of approximation. Figure 1 outlines the relationship between various steps in the calculation of some experimental observables. Central to all other steps is the calculation of the adiabatic electronic wavefunctions for all states of interest. From the wavefunctions one can obtain first order properties and coupling matrix elements for estimating corrections due to coupling of states by non-adiabatic or spin-orbit effects. Methods which by-pass the wavefunction such as Xα or density functional models (12) are not yet sufficiently general to treat this wide class of chemical problems.

Each box in Figure 1 represents its own peculiar computing problems. The algorithms for various steps are at various levels of sophistication depending on the relative cost, difficulty, and interest in the results. The initial calculation of electronic wavefunctions and energy surfaces have preoccupied quantum chemists for thirty years. The calculation of adiabatic scattering and reaction rates has received much attention in recent years (13). The accurate calculation of vibrational-rotational levels is nearly as difficult but has received little attention until very recently. Equally accurate formalisms in the coupled state model do not exist because no general algorithmetric formalism exists for handling the electronic part of the problem. No vibrational-rotational spectrum has yet been computed from an ab initio approach taking full account of Born-Oppenheimer coupling in a Jahn-Teller-Renner situation. Generally speaking the whole area of coupled electronic state calculations lacks a workable algorithm. First order perturbation theory, while suggestive, is often not a quantitative tool.

The rest of this paper will deal exclusively with algorithms for construction of electronic wavefunctions because these are central to the overall problem. In order to appreciate the methods used, one must recall that we are interested in solving a partial differential equation eigenvalue problem for several wavefunctions at several different arrangements of the nuclei. This differential equation involves one- and two-body operators in the potential energy operator and partial derivatives with respect to 3N coordinates (where N is the number of electrons).

For benzene, for example, there are 12 nuclei and 42 electrons. The reasonable aspiration of finding the equilibrium geometry and force constants for the first 10 states would involve solving a partial differential

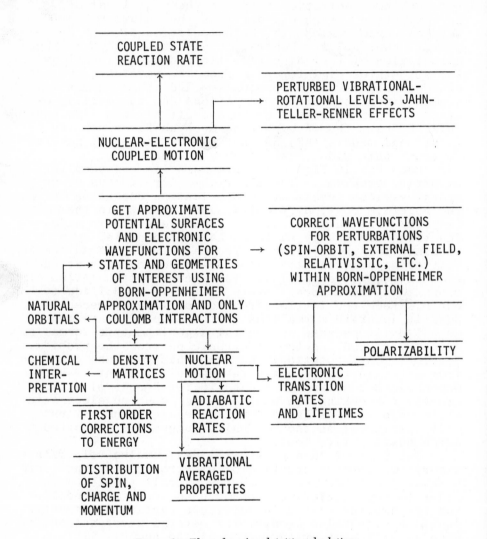

Figure 1.   *Flow chart for ab initio calculations*

equation of this type in 126 independent variables. The only reason it is possible here is that (1) the fixed field due to the nuclei dominates over the electron-electron repulsion so the electronic motions are usually not strongly coupled to each other, (2) it is impossible for a large collection of mutually-repulsive particles to avoid each other if they are constrained to remain in the same region of space, and (3) electrons are indistinguishable so the coordinates are permutational equivalent. Hence the antisymmetric independent particle approximation which leads to a pseudo-separation of variables is often a good first approximation.

Now consider the resources available for solving this (or a similar) problem if some government agency decides these results are vital to the national welfare. It would then be possible to spend up to 10^4 hours of CDC7600 time on this problem (about $10,000,000). This will allow about 10^{14} arithmetic operations (addition or multiplication). Also we can assume that at most 10^6 words of high speed core memory, 10^7 words of low speed core, 10^8 words of disk or drum storage, and 10^9 words of sequential tape storage are available. By present standards this would be a very large calculation since every member given here is a factor of 10^3 larger than what is typically used.

If one wavefunction at one set of nuclear coordinates were sought by numerical integration using only two points in each coordinate, a grid of $2^{126} \simeq 10^{38}$ points would be required. If spin and antisymmetry are taken into account the situation is even worse. Since no two electrons can be at the same point with the same spin at least N positions must be considered for each electron and the minimum grid contains $42! = 10^{51}$ points in 3N space.

The only method found so far which is flexible enough to yield ground and excited state wavefunctions, transition rates and other properties is based on expanding all wavefunctions and operators in a finite discrete set of basis functions. That is, a set of one-particle spin-orbitals $\{\phi_j\}_{j=1}^D$ are selected and the wavefunction is expanded in Slater determinants based on these orbitals. A direct expansion would require writing Ψ as

$$\Psi = \sum_I C_I \Phi_I$$

$$\Phi_I = \det(\phi_{i_1}, \phi_{i_2}, \dots \phi_{i_N}) \qquad 1 \le i_1 \le_2 \le \dots \quad \le_{i_N} \le D$$

Since the number of possible Slater determinants is $\binom{D}{N}$, this again gives an exponential dependence on N. For example, the simplest chemically reasonable orbital basis set for benzene has 72 spin orbitals and $\binom{72}{42} \approx 10^{21}$. Clearly this expansion method is feasible only if very few of the Slater determinants actually contribute to each of the first few wavefunctions. Hence a method is required for constructing the orbitals ϕ_i so that it is known in advance that relatively few of the Φ_I will be important.

The standard method for selecting the Φ_I is to ask for the ϕ_i which maximize the importance of one or more terms in the sum. This gives the self-consistent-field (SCF) or multiconfiguration SCF (MC-SCF) equations. If each ϕ_i is expanded as a linear combination of some fixed set of basis functions $\{f_j\}_{j=1}^{d}$ the coefficients can be found by an extension of the Roothaan SCF equations.

Figure 2 gives an outline of the steps in this approach along with the cost (in machine operations) of each step. For benzene this still requires about 10^8 operations to form all the integrals required to represent the energy operators, in the simplest reasonable basis set (d=36), 10^7 operations to find one SCF wavefunction, 10^8 operations to form the integrals over molecular orbitals and about 10^8 operations to obtain a good expansion for the wavefunction. If 10 wavefunctions were wanted at 10^3 nuclear arrangements the total cost would approach 10^{13} operations. Further, if a good basis set were used including Rydberg orbitals which are known to be important for some of the lowest excited states the number of basis functions could easily be quadrupled and the number of arithmetic operations would be very nearly 10^{15}. In this example the storage available would present no problem although all of the integrals would not fit into high speed core at one time.

In the following sections of this paper some of the algorithms involved in the various steps shown in Figure 2 are presented in detail. Emphasis is placed on concepts which might be useful outside of quantum chemistry. From the previous discussion it should be clear, however, that ab initio calculations are inherently expensive. Since few research projects can afford to use more than 10^{11} arithmetic operations or 10^7 words of memory (of all sorts) only relatively small molecules can be treated in detail. For medium size molecules one must be content with SCF calculations at only a few nuclear arrangements. For very large mole-

ARITHMETIC
OPERATIONS

SELECT
BASIS
SET $\{f_i\}_{i=1}^{d}$

FORM INTEGRALS
$<f_i|h|f_j>$

$[f_if_j||f_kf_\ell]$
$100d^4$ to $5\times10^5d^2$

FORM SCF ORBITALS
$\phi_i = \Sigma a_{ji}f_i$
$20d^4$ to $300d^3$

FORM INTEGRALS
$<\phi_i|h|\phi_j>$

$[\phi_i\phi_j||\phi_k\phi_e]$
d^5 (or N^2d^2)

SELECT CONFIGURATIONS
by perturbation theory or
other rules
keep K configurations
$100N^2 (d-\frac{N}{2})^2$

FORM CI MATRIX
FORMULA
$250\ K^2/N^2$

ELEMENT
$25\ K^2/N^2$

FIND EIGENVECTOR
AND ENERGY
$25\ K^2/N^2$

FORM DENSITY
& MOLEC PROP.
$100[K^2/N^2(d-\frac{N}{2})]$ or $50d^3$

Figure 2.   Unit operations in calculating a wavefunction

cules (more than 500 valence electrons) in the absence
of symmetry, even the crudest calculation becomes
excessively expensive.

Integral Calculation

The integrals involved in typical quantum chemical
calculations are of the form ($\underline{17},\underline{18}$)

$$B_{ij} = \int f_i^*(\underline{r}) \, B \, f_j(\underline{r})d\tau$$

and

$$Q_{ijk\ell} = \int f_i^*(\underline{r}_1)f_j(\underline{r}_1)Q(\underline{r}_{12})f_k(\underline{r}_2)f_\ell(\underline{r}_2)d\tau_1 d\tau_2$$

where B is ∇^2, ∇, r, $r:r$, r^{-1}, Y_{1m}/r^2, Y_{2m}/r^3, etc.
and Q is r_{12}^{-1}, $Y_{1m}(\Omega_{12})/r_{12}^3$, etc.

The basis functions f_i must therefore be chosen as a
compromise between the best representation of the wavefunction (which requires the fewest f_i and hence fewest
integrals) and the easiest functions to integrate. For
atoms, Slater orbitals, $r^n Y_{\ell m}(\Omega)$, and numerical orbitals, $R(r)Y_{\ell m}(\Omega)$ with R given numerically, are sufficiently accurate and simple. For diatomics, Slater
orbitals have remained the best choice because the integrals can be done with reasonable effort. Polyatomic calculations, however, were blocked for many
years because of the difficulty of evaluating electron
repulsion (r_{12}^{-1}) integrals with Slater orbitals. It has
been known for some time that gaussian orbitals, $x^n y^\ell z^m$
$\exp(-\alpha r^2)$, have certain peculiar properties which make
the integrals relatively easy to obtain ($\underline{14}$). On the
other hand this functional form is not much like the
wavefunction of a coulomb potential so more functions
are required.

In recent years a compromise has been found which
presently dominates polyatomic calculations. Each
function f_i is expanded as a linear combination of
gaussian orbitals (f is then called a contracted gaussian function). Since this is basically a numerical
fitting procedure, various choices have been suggested
for the contraction scheme. The most popular choices
are presently Pople's approximations ($\underline{15}$) to Slater
orbitals and Dunning's approximations ($\overline{\underline{16}}$) to free atom
Hartree-Fock orbitals.

Because they are the most difficult and most numerous of the integrals routinely needed, let us consider
the electron repulsion integrals

$$[ij||k\ell] = \int g_i^*(\underline{r}_1)g_j(\underline{r}_1)\ r_{12}^{-1}g_k(\underline{r}_2)g_\ell(\underline{r}_2)d\tau_1 d\tau_2$$

in more detail for the case that all of the g_i are simple normalized gaussian "lobes"

$$g_1(\underline{r}) = N_i f_i(\underline{r})$$

$$f_i = \exp(-\alpha_i|\underline{r}-\underline{R}_i|^2)$$

$$N_i = (2\alpha_i/\pi)^{3/4}$$

centered at positions $\underline{R}_i$ respectively. This is a "four-center" integral if all the positions are different and is extremely difficult to evaluate using any other type of basis functions. For gaussians, however

$$g_i(\underline{r}_1)g_j(\underline{r}_1) = K_{ij}f_p(\underline{r}_1)$$

where

$$\underline{R}_p = (\alpha_i\underline{R}_i + \alpha_j\underline{R}_j)/(\alpha_i + \alpha_j)$$

$$\alpha_p = \alpha_i + \alpha_j$$

$$f_p = \exp(-\alpha_p|\underline{r}-\underline{R}_p|^2)$$

and

$$K_{ij} = N_iN_j\exp(-\alpha_{ij}|\underline{R}_i-\underline{R}_j|)^2$$

$$\alpha_{ij} = \alpha_i\alpha_j/(\alpha_i+\alpha_j)$$

so the integral reduces easily to a two center integral

$$[ij||k\ell] = K_{ij}K_{k\ell}\int f_p(\underline{r}_1)r_{12}^{-1}f_q(\underline{r}_2)d\tau_1 d\tau_2.$$

This may be further simplified by the change of variables

$$\underline{r} = \tfrac{1}{2}(\underline{r}_1+\underline{r}_2),\ \underline{r}_{12} = \underline{r}_1-\underline{r}_2 \text{ to obtain}$$

$$f_p(\underline{r}_1)f_q(\underline{r}_2) = f_s(\underline{r})f_t(\underline{r}_{12})$$

where

$$\alpha_s = \alpha_p + \alpha_q$$

$$R_s = [\alpha_p(\tfrac{1}{2}\underline{r}_{12} - \underline{R}_p) + \alpha_q(-\tfrac{1}{2}\underline{r}_{12} - R_q)]/\alpha_s$$

and

$$\alpha_t = \alpha_p \alpha_q / (\alpha_p + \alpha_q)$$

$$R_t = \underline{R}_p - \underline{R}_q$$

Hence

$$[ij||k\ell] = K_{ij}K_{k\ell}\int d\tau_{12} f_t(r_{12})r_{12}^{-1}\int d\tau \; f_s(\underline{r})$$

The $\underline{r}$ integration can now be done to give

$$\int d\tau \; f_s(\underline{r}) = (\pi/\alpha_s)^{3/2}$$

Since this is independent of $\underline{R}_s$ (and hence of $\underline{r}_{12}$), one is left with

$$[ij||k\ell] = K_{ij}K_{k\ell}(\pi/\alpha_s)^{3/2}\int d\tau_{12} f_t(r_{12})r_{12}^{-1}.$$

The angular integration in this final three dimensional integral is easily done if a spherical coordinate system is introduced with the z axis chosen along $\underline{R}_t$:

$$\int d\tau_{12} f_t(r_{12})r_{12}^{-1} = \int_0^\infty dr_{12}r_{12}\int_0^\pi \sin\theta d\theta \int_0^{2\pi} d\phi \exp(-\alpha_t r_{12}^2)$$

$$\times \; \exp(-\alpha_t R_t^2)\exp(-2\alpha_t r_{12}R_t\cos\theta)$$

$$= \frac{\pi}{\alpha_t R_t}\int_0^\infty dr_{12}[\exp(-\alpha_t|r_{12}-R_t|^2)$$

$$-\exp(-\alpha_t|r_{12}+R_t|^2)]$$

$$= 2\pi\alpha_t^{-3/2}R_t^{-1}\int_0^{R_t\alpha_t^{\frac{1}{2}}} dr \; \exp(-r^2).$$

The remaining integral is closely related to the error function

$$\text{erf}(t) = 2\pi^{-\frac{1}{2}} \int_0^t dr \, \exp(-r^2).$$

Because this expression for $[ij||k\ell]$ reduces to $0/0$ for $R_t = 0$, it is customary to define a related auxilary function

$$F_0(T) = T^{-\frac{1}{2}} \int_0^{T^{\frac{1}{2}}} dr \, \exp(-r^2)$$

so

$$[ij||k\ell] = \frac{2\pi^{5/2}}{\alpha_t \alpha_s^{3/2}} K_{ij} K_{k\ell} F_0(T)$$

if

$$T = \alpha_t R_t^2.$$

If the overlap charge

$$S_{ij} = \int g_i(\underline{r}_1) g_j(\underline{r}_1) d\tau_1$$

$$= (\pi/\alpha_p)^{3/2} K_{ij}$$

is introduced,

$$[ij||k\ell] = S_{ij} S_{k\ell} \alpha_t^{\frac{1}{2}} 2\pi^{-\frac{1}{2}} F_0(T)$$

$$= S_{ij} S_{k\ell} R_t^{-1} \text{erf}(\sqrt{T}).$$

For large T, $\text{erf}(\sqrt{T}) \to 1$ and the formula for $[ij||k\ell]$ reduces to that for two charges of magnitude S_{ij} and $S_{k\ell}$ interacting at a distance R_t. For small T, $F_0(T) \to 1$ and $[ij||k\ell]$ corresponds to the overlap charges interacting at an average distance of $(\pi/4\alpha_t)^{\frac{1}{2}}$. For all T, $F_0(T) \le 1$ so

$$[ij||k\ell]/S_{ij}S_{k\ell} \le 2\pi^{-\frac{1}{2}} \alpha_t^{\frac{1}{2}}.$$

Since $S_{k\ell} \le 1$ and $\alpha_t < \alpha_p$,

$$0 \le [ij||k\ell] \le 2\pi^{-\frac{1}{2}} \alpha_p^{\frac{1}{2}} S_{ij}.$$

Contracted gaussian lobes (i.e. combinations of only simple gaussians) are frequently used as basis functions (21). For large molecules the lobes may be

centered in widely scattered parts of the molecule so
that most of the S_{ij} overlap charges are quite small
($\leq 10^{-13}$). The energy and wavefunction seem to depend
only on fixed point accuracy in the integrals [i.e.,
$\pm 10^{-6}$ absolute (not relative) error in each integral
gives about $\pm 10^{-6}$ absolute error in energy]. Hence
most integrals do not need to be evaluated for large
molecules. Further, many of the integrals can be
eliminated by a test based only on one charge distribution. Thus, although $\sim d^4$ integrals need to be done
for small molecules, only $\sim d^2$ integrals are needed for
large molecules.

   Those integrals which remain to be done can be
written so they involve one exponential, one square
root, and either $F_0(T)$ or $erf(\sqrt{T})$. Each of these three
functions involve about the same amount of time although
the square root can be made 30% faster than the standard square root routine furnished with the computer
software package. Since billions of these basic
[ij||kℓ] integrals must be evaluated in a typical large
calculation, it is essential that the fastest possible
algorithm be used. In this regard it is best to evaluate $F_0(T)$ for small T and $erf(\sqrt{T})$ for large T. By
judicious choice of intervals a short Chebyshev series
for $F_0(T)$ or $erf(\sqrt{T})$ can be found on each interval ([19],
[20]). Although this involves storing about 4000 coefficients and pointers, the resulting algorithm is nearly
twice as fast as one based on larger intervals and
longer series or on a Taylor series for short intervals.
This division into intervals is simplified by the fact
that only $0 \leq T < 30$ need be considered since $erf(\sqrt{30})$
is one to twelve significant figures.

   This analysis is typical of the approach to electron repulsion integrals. Use of cartesian gaussian
functions gives rise to a more general basic integral
([17])

$$F_n(T) = \int_0^1 e^{-Tu^2} u^{2n}\, du$$

Similarly, Slater orbitals for diatomic molecules give
integrals of the form ([22])

$$A_{nq}(\alpha) = \int_1^\infty e^{-\alpha u} u^n (u^2-1)^q\, du$$

and ([23])

$$B_{nq\ell m}(\beta) = \int_{-1}^1 e^{-\beta u} u^n P_\ell^m(u) (1-u^2)^{m/2+q}\, du$$

Rather elaborate recursion relations can be found for
all these integrals when care is taken to preserve
numerical accuracy. Since usually all values of n are
needed anyway, the intermediate values of n as well as
the largest value n=N and the smallest n=0 are useful.
For example, the recursion relation

$$(2n+1)F_n(T) = 2T\ F_{n+1}(T) + e^{-T}$$

is stable for recurring downward on n but is unstable
for recurring upwards (for small T/n) because
$(2n+1)F_n(T) \approx e^{-T}$.

  Consequently, evaluation of $F_n(T)$ involves different
schemes depending on the value of N and T. For $T \gtrsim N$,
upward recurrence from F_0 is possible without loss of
significant figures. For T < N, downward recurrence
must be used starting from $F_N(T)$. For most functions
this situation would require either a set of tables for
every possible starting value of N or else one table
for an N* greater than any N which can occur followed
by downward recurrence from N*. The particular function
F dealt with here, however, obeys the relationship

$$\frac{d}{dT}\ F_n(T) = -F_{n+1}(T)$$

so the Taylor series has the simple form

$$F_n(T) = \sum_{k=0}^{\infty} F_{n+k}(T_o)\ (T_o-T)^k/k!$$

The convergence rate of this series is nearly independent of n ($F_{n+k+1}/F_{n+k} \approx 1$ for small T) so a table of
$F_n(T_o)$ at a sequence of intervals of T_o for n from zero
N*+K (an interval width of 0.1 requires a K of 6 for
twelve significant figures) suffices for all values of
N and T. As for F_o, at large T it is better to evaluate
a generalized error function

$$G_n(\sqrt{T}) = \int_o^{\sqrt{T}} e^{-u^2} u^{2n} du$$

$$G_{n+1}(\sqrt{T}) = (n+\tfrac{1}{2})G_n(\sqrt{T}) - T^{n+\frac{1}{2}}e^{-T}$$

Hence an efficient algorithm must recognize several
ranges of T:

$$T = 0 \qquad\qquad F_n = (2n+1)^{-1}$$

$$0 < T \leq N \qquad\qquad F_N \text{ by Taylor series,} \\ \text{recur down}$$

$$N < T \leq T* \qquad\qquad F_O \text{ by Chebyshev series,} \\ \text{recur up}$$

$$\max (N, T*) < t < T** \qquad G_O \text{ by Chebyshev series,} \\ \text{recur up}$$

$$T** < T < T*** \qquad\qquad G_O = 1, \text{ recur up}$$

$$T*** < T \qquad\qquad G_O = 1, \; G_{n+1} = (n+\tfrac{1}{2})G_n$$

where $T* \approx 7$, $T** \approx 30$, $T*** \approx 30+3N$ if 13 figure accuracy is wanted.

Self-Consistent-Field

The simplest approximate wavefunction for an open-shell molecule is the spin-unrestricted Hartree-Fock function

$$\psi = (N!)^{-\frac{1}{2}} \det\{\phi_1 \phi_2 \cdots \phi_k \overline{\phi}_{k+1} \cdots \overline{\phi}_N\}$$

where N is the number of electrons and

$$\phi_i = \sum_{j=1}^{d} c_{ji} f_j(\underline{r})\alpha$$

$$\overline{\phi}_i = \sum_{j=1}^{d} c_{ji} f_j(\underline{r})\beta$$

are orthonormal spin-orbitals. The expectation value of the energy, $\langle\psi|H|\psi\rangle$, is a quartic polynomial $E(\underline{c})$ in the Nd variables $\underline{c}$. The orthonormality constraints form a set of subsidiary quadratic constraints of the form

$$G_i(\underline{c}) = 0 \qquad i = 1 \cdots L$$

The self-consistent-field algorithm is an iterative method for finding the coefficients $\underline{c}$ which minimize $E(\underline{c})$ subject to these constraints.

This algorithm may be derived from the Euler-Lagrange equations

$$\partial E/\partial c_{ji} = \sum_k \lambda_k \, \partial G_k/\partial c_{ji}$$

which are cubic in $\underline{c}$. The wavefunction ψ is unchanged by a unitary transformation among the spin-up or spin-down orbitals. Roothann ($\underline{24}$) has shown how this arbitrariness may be used to change the Euler-Langrange equations to the pseudo-eigenvalue form

$$\underline{F}(\underline{c})\underline{c}_k = \varepsilon_k \underline{S}\underline{c}_k$$

where $\underline{F}$ is a quadratic polynomial in the $\underline{c}$ coefficients (which is still somewhat arbitrary). Since this cubic equation cannot be solved explicitly, one can attempt an iterative solution in the form

$$\underline{F}(\underline{c}^{(n-1)})\underline{c}_k^{(n)} = \varepsilon_k \underline{S}\underline{c}_k^{(n)}.$$

Although this equation is usually stated as the basis of the iterative algorithm, it often does not lead to rapid convergence ($\underline{25}$). Consequently the $\underline{F}$ matrix is usually modified in four different ways.

(1)   the arbitrariness ($\underline{26}$) in the definition of F is used to ensure that the correction $\delta\underline{c}$ to $\underline{c}$ agrees with the Newton-Raphson solution of the Euler-Lagrange equations to first order in $\delta\underline{c}$.

(2)   the elements of F are extrapolated ($\underline{27}$) from $F(\underline{c}^{(n-3)})$, $F(\underline{c}^{(n-2)})$, and $F(\underline{c}^{(n-1)})$ assuming each element converges geometrically to give $F^{(n-1)}$.

(3)   oscillations are damped by averaging ($\underline{27}$), with appropriate weights $F^{(n-2)}$ and $F(\underline{c}^{(n-1)})$ to give $F^{(n-1)}$.

(4)   oscillations are damped by adding ($\underline{26}$) to $F(\underline{c}^{(n-1)})$ a root-shift $\sum \alpha_j \underline{c}_j^{(n-1)} \underline{c}_j^{(n-1)T}$ to obtain $F(\bar{n}-1)$. The last three of these modifications have the property that $F^{(n)}$ converges to $F(\underline{c})$ as $\underline{c}^{(n)}$ converges to $\underline{c}$; so at convergence the cubic equation is solved.

These methods for controlling convergence of an iterative solution to a complicated set of equations have wide applicability. The extrapolation and damping methods are based on well-known ideas for single variables while root-shifting may be a novel development by quantum chemists.

Spin-restricted and multi-configuration self-consistent-field methods differ in the assumed func-

tional form for ψ. The basic method for solving the
resulting cubic Euler-Lagrange equations remains similar to that just discussed.

Configuration Interaction

Configuration interaction has come to mean any
expansion of the wavefunction in a finite series of N-
electron functions (28)

$$\Psi = \sum C_I \Phi_I(1,\ldots,N)$$

where the C_I satisfy the matrix eigenvalue equations

$$\underline{H}\ \underline{C} = E\ \underline{S}\ \underline{C}$$

$$H_{IJ} = <\Phi_I|H|\Phi_J>$$

$$S_{IJ} = <\Phi_I|\Phi_J>$$

Most CI calculations involve configurations formed
from a common set of orthonormal orbitals by spin and
symmetry adaptation of Slater determinants. In this
case $\underline{S}$ is a unit matrix and the formation of $\underline{H}$ is
greatly simplified.

In most CI calculations the H_{IJ} are first expressed
in terms of basic integrals I_k over orthonormal molecular orbitals as

$$H_{ij} = \sum \Gamma_k{}^{IJ} I_k$$

where the $\Gamma_k{}^{IJ}$ are integral independent coefficients
which constitute a "formula" for H_{IJ}. Generating the
$\Gamma_k{}^{IJ}$ is the most time consuming part of the formation
of $\underline{H}$. Since the $\Gamma_k{}^{IJ}$ are dependent only on the indices
of the orbitals involved in Φ_I and Φ_J they may be used
for several arrangements of the molecular nuclei (as
long as the labels involved in each configuration
remain unchanged).

If Ψ is predominantly one Slater determinant, the
coefficients $\underline{C}$ may be found by many-body-perturbation
theory (29). This theory provides an elegant scheme
for simplifying the perturbation formulas by combining
terms referring to the same I_k integrals.

In the more general case, Ψ involves several
Slater determinants with large coefficients and corresponds to an excited state. In this case no simplified
theory is possible and $\underline{H}$ must be constructed.

The first step in constructing $\underline{H}$ is producing the
list of configurations to be included. At a moderate
level of accuracy only the SCF configuration and other
configurations nearly degenerate with it need be
considered. For higher accuracy more configurations
are needed. These configurations may be classified as
singly, doubly, triply,..excited depending on the
least number of excitations required to form the configuration from one of the dominant ones. For fixed
relative error in the excitation energy of a hydrocarbon
molecule the number of spin orbitals increases in pro-
portion to the number of electrons, N. The number of
k-fold excitations from any one Slater determinant is
then proportional to N^{2k}. If all configurations are
used to all excitation levels there are $\sim N^4$ non-zero
entries in each row of $\underline{H}$ and about λ^N rows (where λ is
a fixed number for fixed relative error and is about
10 for a double zeta basis set).

As noted before, such a large rate of growth with
N cannot be tolerated. Consequently most CI calculations are run with limited excitation levels (typically
only single and double excitations). It is easily
domonstrated, however, that this procedure leads to
increasing error as the number of electrons increases.
In fact, for tightly localized electron pairs, the dominant excitation level is the value of k nearest $\sim 0.01N$
(i.e., for about 200 electrons the double excitations
in aggregate are more important than the SCF configuration and for 400 electrons quadruple excitations should
dominate). Even for molecules with only 40 electrons
quadruple and higher excitations must be considered in
order to reproduce excitation energies (30) or potential
surfaces to an accuracy of ± 0.1 eV. Thus, configuration
interaction calculations for very large molecules are
hopeless unless perturbation theory can be used to
correct for unlinked cluster effects.

For this reason, modern CI calculations are really
limited to high accuracy calculations on small molecules. With this limitation both excited and ground
states may be treated with uniform accuracy provided
the same procedure is followed for each state. This
requires a separate SCF calculation, integral transformation, and CI calculation for each desired state.

Because of the large number of configurations
which can be constructed even with just double excitations, some attention must be paid to limiting the
number which are important. This can be done by constructing molecular orbitals which maximize the convergence rate of the CI series. Ordinary SCF orbitals
offer a reasonable starting set of occupied orbitals
(although localized orbitals may be better). The SCF
virtual orbitals can be improved, however, by use of
approximate natural orbitals (31). These orbitals are
distinguished by the fact that they are largest in the
regions where the wavefunction error is largest. In
terms of such localized corrections only a few double
excitations from each of the electron pairs are required
for reasonable accuracy.

The actual algorithm for evaluating H_{IJ} varies
greatly between different research groups. The crudest,
but most general, approach is to assume each configuration is formed as a short sum of Slater determinants

$$\Phi_I = \sum_\nu X_{\nu,I} \ \det(\phi_{\nu 1}, \phi_{\nu 2} \cdots \phi_{\nu N})$$

which produces a spin-eigenfunction from orthonormal
spin-restricted spin-orbitals (i.e., the spin-up and
spin-down spin-orbitals occur in pairs which differ
only in spin). Then H_{IJ} is zero if all of the Slater
determinants in Φ_I differ by at least three substitu-
tions from all of the determinants in Φ_J. Since most
matrix elements are zero, a rapid test for this condition is essential. Usually a configuration is specified by the list of space-orbitals (spin-independent)
which occur in every Slater determinant in the configuration. These space orbital occupations are specified by two binary words where each bit is on or off in
one word depending on whether the corresponding orbital
is singly occupied or not and on or off in the other
word depending on whether the corresponding orbital is
doubly occupied. Boolean arithmetic on these words can
easily produce a word which indicates which occupations
have changed and the bit count of this word can give the
number of changes. For those H_{IJ} which have to be
evaluated, there are different formulae depending on the
number of orbitals by which Φ_I and Φ_J differ (28,32).

Matrix Manipulations

Storage. One of the more serious computational
problems in quantum chemistry is the storage, manipulation, and retrieval of large arrays of real numbers.
If some care is not taken, a calculation may be need-
lessly limited by the storage capacity of central
memory, disks, or tapes.

The largest arrays which occur in calculations are
of two types. One arises from the electron repulsion
integrals and grows in size like the fourth power of the
number of basis functions. The other is the configuration interaction hamiltonian matrix which grows like
the square of the number of configurations. Many other
smaller arrays, whose size is proportional to the square
of the number of basis functions, occur throughout the
calculation.

For non-symmetric matrices of dimension nxm with
few zero entries the most efficient storage pattern is
rectangular (hereafter referred to as R) with the
location of the i,j element computed as $L = i+n(j-1)$.
For real symmetric matrices of dimension n, a triangular pattern (referred to as T) is used with the location
of i,j computed as $L = i+j(j-1)/2$ for $i \leq j$ [or
$L = j+i(i-1)/2$ for $i \geq j$]. The CI hamiltonian matrix is
a large real symmetric matrix with mostly zero entries
(provided orthonormal configurations constructed from
orthonormal orbitals are used). If more than half the
entries are zero it is more efficient to omit zero
entries and include the index as a label (if the word
length is long and the matrix is small enough, this
label may be packed into the insignificant bits of the
matrix element).

The electron repulsion integrals are more complicated to store if point group symmetry is used to
reduce their number. In general the integrals may be
classified into blocks depending on the symmetry of the
four orbitals involved in the integral $[i_1 i_2 || i_3 i_4]$.
Integrals from the block labeled with symmetries
$\Gamma_1, \Gamma_2, \Gamma_3, \Gamma_4$ can be stored in six different patterns:
RRR, RTR, TTR, RTT, RRT, and TTT where the first
letter tells whether a rectangular ($\Gamma_1 \neq \Gamma_2$) or triangular
($\Gamma_1 = \Gamma_2$) pattern is used to compute the first charge
distribution location L_{12}. The second letter indicates
whether a rectangular ($\Gamma_3 \neq \Gamma_4$) or triangular ($\Gamma_3 = \Gamma_4$)
pattern is used to compute the second charge distribu-
tion location L_{34} and the final letter indicates whether
a rectangular ($\Gamma_1 \neq \Gamma_3$ or $\Gamma_2 \neq \Gamma_4$) or triangular ($\Gamma_1 = \Gamma_3$ and
$\Gamma_2 = \Gamma_4$) pattern is used to compute the integral location

L_{1234}. Zero blocks are omitted, of course, and it is sufficient to consider $\Gamma_1 \geq \Gamma_2$, $\Gamma_3 \geq \Gamma_4$, and $\Gamma_1(\Gamma_1-1)/2 + \Gamma_2 \geq \Gamma_3(\Gamma_3-1)/2 + \Gamma_4$. Non-zero integrals over symmetry orbitals or molecular orbitals are usually not small so no further simplification is possible. Non-zero integrals over atomic basis functions may be quite small, however, and large numbers of these can be omitted if labels are retained.

<u>Transformations</u>. A frequently occurring step in calculations is a change of basis via a linear transformation. That is, a new set of basis functions (such as molecular orbitals, group orbitals, natural orbitals, etc.) are defined as linear combinations of the original atomic orbitals, by

$$g_i(\underline{r}) = \sum_{j=1}^{d} w_{ji} f_j(\underline{r}), \quad i=1\ldots d' \leq d.$$

Matrix elements of one-body hermitian operators (such as kinetic energy, nuclear attraction, the Fock operator, etc.) have the form

$$B_{ij} = \int f_i(\underline{r})^* B f_j(\underline{r}) d\tau$$

in terms of the original basis functions. The new matrix elements

$$\overline{B}_{ij} = \int g_i(\underline{r})^* B g_i(\underline{r}) d\tau$$

are easily computed from the B_{ij} by the matrix transformation

$$\overline{\underline{B}} = \underline{w}^\dagger \underline{B} \underline{w}.$$

If symmetry is considered, one may also encounter unsymmetrical blocks of the B matrix defined by

$$B_{ij}^{\Gamma_i, \Gamma_2} = \int f_{i,\Gamma_1}(\underline{r})^* B \, f_{j,\Gamma_2}(\underline{r}) d\tau$$

where f_i, Γ_1 is the ith function in symmetry block Γ_1. In this case there will be a different $W_{(\Gamma)}$ matrix for each symmetry block and one must compute all non-vanishing matrices of the form

$$\overline{\underline{B}}_{(\Gamma_1,\Gamma_2)} = \underline{W}^\dagger{}_{(\Gamma_1)}\underline{B}_{(\Gamma_1,\Gamma_2)}\underline{W}_{(\Gamma_2)}.$$

Thus, generally, two matrix transformation algorithms are required, one for $\underline{B}$ stored triangularly ($\Gamma_1=\Gamma_2$) and one for $\underline{B}$ stored rectangularly ($\Gamma_1\neq\Gamma_2$). The transformation could be written as a double sum

$$\overline{B}_{ij} = \sum_{k\ell}^{d_1,d_2} W^*{}_{(\Gamma_1)ki}B_{(\Gamma_1,\Gamma_2)k,\ell}\,W_{(\Gamma_2)\ell j} \qquad i=1\dots\overline{d}_1,j=1\dots\overline{d}_2$$

Direct evaluation in this form requires $d_1d_2\overline{d}_1\overline{d}_2$ multiplication. On the other hand the multiplication $\underline{Y} = \underline{B}_{(\Gamma_1\Gamma_2)}\underline{W}_{(\Gamma_2)}$ followed by $W^T{}_{(\Gamma_2)}Y$ requires only $d_1d_2\overline{d}_2+\overline{d}_1d_1\overline{d}_2$ multiplications (or $d_1d_1\overline{d}_2 + \overline{d}_1d_2\overline{d}_2$ if the multiplications are done in the opposite order).

Figures 3,4 show an outline of algorithms for the triangular and rectangular cases for matrices small enough to fit entirely into high speed core. These algorithms are designed with one additional principle in mind. Namely, the only real variation between different ways of doing matrix multiplication is the cost of indexing and amount of scratch storage used. Double subscripts should usually be avoided and as far as possible matrix elements should be accessed sequentially. For this reason it is best to carry out the rectangular transformation as $\underline{Y} = \underline{B}^T{}_{\Gamma_1\Gamma_2}\underline{W}_{(\Gamma_1)}$ followed by $\overline{\underline{B}}_{(\Gamma_1\Gamma_2)} = \underline{Y}^T\underline{W}_{(\Gamma_2)}$. Scratch storage is reduced by using each column of $\underline{Y}$ as soon as it is formed to do the second multiplication. The triangular transformation is furthur complicated by the fact that both $\underline{B}$ and $\overline{B}$ are stored in a triangular pattern which increases the complexity of indexing.

Transformation of the two electron integrals is a much more time consuming step. If $R(i_1,i_2,i_3,i_4)$ is the integral

$$R(i_1,i_2,i_3,i_4)=\int f_{i_1}(\underline{r}_1)^* f_{i_2}(\underline{r}_1)r_{12}^{-1}f_{i_3}(\underline{r}_2)^* f_{i_4}(\underline{r}_2)d\tau_1 d\tau_2$$

and $\overline{R}$ is the transformed integral

$$\overline{R}(i_1,i_2,i_3,i_4)=\int g_{i_1}(\underline{r}_1)^* g_{i_2}(\underline{r}_1)r_{12}^{-1}g_{i_3}(\underline{r}_2)^* g_{i_4}(\underline{r}_2)d\tau_1 d\tau_2$$

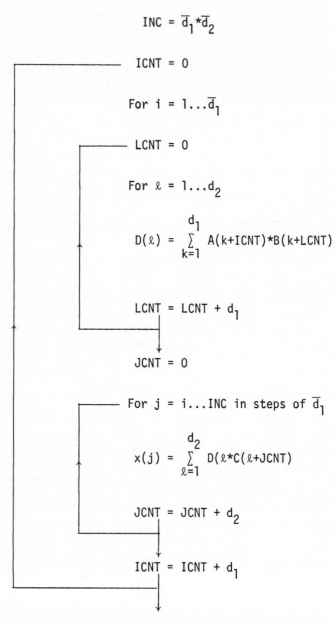

$$INC = \overline{d}_1 {}^* \overline{d}_2$$

$$ICNT = 0$$

$$For\ i = 1...\overline{d}_1$$

$$LCNT = 0$$

$$For\ \ell = 1...d_2$$

$$D(\ell) = \sum_{k=1}^{d_1} A(k+ICNT)*B(k+LCNT)$$

$$LCNT = LCNT + d_1$$

$$JCNT = 0$$

$$For\ j = i...INC\ in\ steps\ of\ \overline{d}_1$$

$$x(j) = \sum_{\ell=1}^{d_2} D(\ell*C(\ell+JCNT)$$

$$JCNT = JCNT + d_2$$

$$ICNT = ICNT + d_1$$

Figure 3. Transformation of a real non-symmetric matrix, X =
$$A^T BC$$

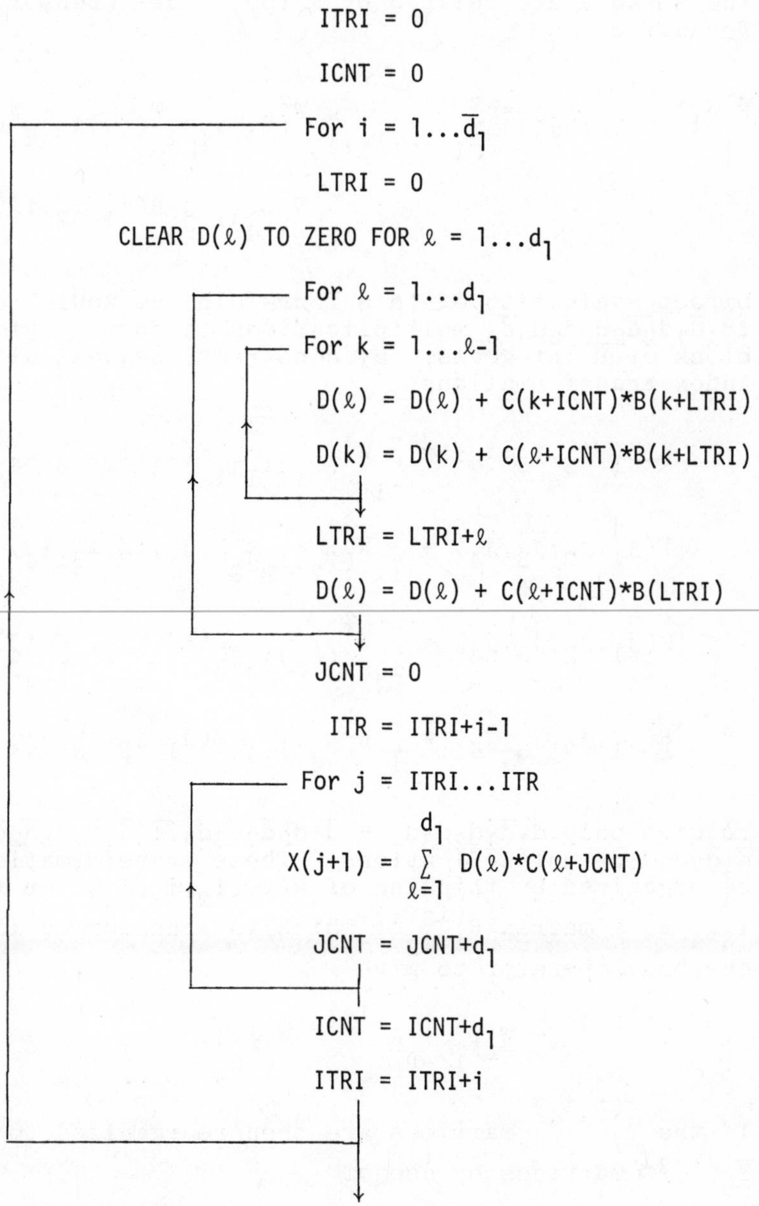

$$ITRI = 0$$

$$ICNT = 0$$

For $i = 1 \ldots \overline{d}_1$

$$LTRI = 0$$

CLEAR $D(\ell)$ TO ZERO FOR $\ell = 1 \ldots d_1$

For $\ell = 1 \ldots d_1$

For $k = 1 \ldots \ell-1$

$$D(\ell) = D(\ell) + C(k+ICNT)*B(k+LTRI)$$

$$D(k) = D(k) + C(\ell+ICNT)*B(k+LTRI)$$

$$LTRI = LTRI+\ell$$

$$D(\ell) = D(\ell) + C(\ell+ICNT)*B(LTRI)$$

$$JCNT = 0$$

$$ITR = ITRI+i-1$$

For $j = ITRI \ldots ITR$

$$X(j+1) = \sum_{\ell=1}^{d_1} D(\ell)*C(\ell+JCNT)$$

$$JCNT = JCNT+d_1$$

$$ICNT = ICNT+d_1$$

$$ITRI = ITRI+i$$

Figure 4. Transformation of a real symmetric matrix, $X = C^T BC$

the $\bar{R}$ and R are related by a four-index (tensor) transformation.

$$\bar{R}(j_1,j_2,j_3,j_4) = \sum_{i_1,i_2,i_3,i_4} W^*(\Gamma_1)_{i_1j_1} W(\Gamma_2)_{i_2j_2} W(\Gamma_3)^*_{i_3j_3}$$

$$W(\Gamma_4)_{i_4j_4} R(i_1,i_2,i_3,i_4)$$

Direct evaluation of this four-fold sum would require $4d_1\bar{d}_1d_2\bar{d}_2d_3\bar{d}_3d_4\bar{d}_4$ multiplications to form a symmetry block of $\bar{R}$ integrals. By constrast, sequential one-index transformations

$$X(j_1,i_2,i_3,i_4) = \sum_{i_1} W^*(\Gamma_1)_{i_1j_1} R(i_1,i_2,i_3,i_4)$$

$$Y(j_1,j_2,j_3,j_4) = \sum_{i_2} W(\Gamma_2)_{i_2j_2} X(j_1,i_2,i_3,i_4)$$

$$Z(j_1,j_2,j_3,j_4) = \sum_{i_j} W^*(\Gamma_3)_{i_3j_3} Y(j_1,j_2,i_3,i_4)$$

$$\bar{R}(j_1,j_2,j_3,j_4) = \sum_{i_4} W(\Gamma_4)_{i_4j_4} Z(j_1,j_2,j_3,j_4)$$

require only $d_1\bar{d}_1d_2d_3d_4 + \bar{d}_1d_2\bar{d}_2d_3d_4 + \bar{d}_1\bar{d}_2d_3\bar{d}_3d_4 + \bar{d}_1\bar{d}_2\bar{d}_3d_4\bar{d}_4$ multiplications. These transformations can be organized by thinking of $R(i_1,i_2,i_3,i_4)$ for fixed i_3i_4 as a matrix $R^{(i_3i_4)}_{i_1i_2}$ which is transformed like a one-body operator to give

$$\underline{Y}^{(i_3i_4)} = \underline{W}(\Gamma_1)^\dagger \underline{R}^{(i_3i_4)} \underline{W}(\Gamma_2).$$

If the $\underline{Y}^{(i_3i_4)}$ matrices are then reorganized to give $\underline{\bar{Y}}^{(j_1j_2)}$ matrices by use of

$$Y(j_1,j_2,i_3,i_4) = Y^{(i_3,i_4)}_{j_1,j_2} = \bar{Y}^{(j_1j_2)}_{i_3,i_4}$$

the $\overline{R}$ integrals can be formed from

$$\overline{\underline{R}}^{(j_1 j_2)} = \underline{W}_{(\Gamma_3)}{}^\dagger \overline{\underline{Y}}^{(j_1 j_2)} \underline{W}_{(\Gamma_4)}$$

$$\overline{R}(j_1, j_2, j_3, j_4) = \overline{R}^{(j_1 j_2)}_{j_3 j_4}$$

The use of six different storage patterns for the two-electron integrals requires six different algorithms for carrying out the transformation. Only the simplest (RRR) will be presented in detail here (33). Since the number of integrals usually exceeds the amount of high speed core available (and usually low speed core as well) a transformation using minimum core will be discussed (assuming disk is large enough to hold one block of R). Suppose the integrals $R(i_1 i_2 i_3 i_4)$ are originally arranged so that $\underline{R}^{(1,1)}$, $\underline{R}^{(2,1)}$... appear in sequential order on a sequential file. The range of $(i_3 i_4)$ can be blocked into $d_3 d_4 / n_{34}$ groups of size n_{34} (with a smaller group at the end if needed). Each group of n_{34} $\underline{R}$ matrices can then be transformed by a standard two subscript transformation to leave n_{34} $\underline{Y}$ matrices in sequential order (in the same space in core originally occupied by the $\underline{R}$ matrices). Storage for the $\underline{W}$ matrices and one scratch region for $\underline{W}^T_{(\Gamma_1)}\underline{R}^{(i_3 i_4)}$ are needed in addition to the space for the $\underline{R}$ arrays. The $j_1 j_2$ subscripts on each $\underline{Y}^{(i_3 i_4)}$ array can also be blocked into $\overline{d}_1 \overline{d}_2 / n_{12}$ blocks of size n_{12} and the $\underline{Y}$ arrays can be written to disk in blocks of size n_{12} by n_{34} as a random access file. When all $\underline{R}$ matrices have been transformed, a block of $\overline{Y}^{(j_1 j_2)}$ matrices is easily formed in core by reading all appropriate pieces from disk. The $\underline{Y}$ arrays can then be transformed by a standard two subscript transformation and written to a sequential file. This method requires $d_1 d_2 n_{34}$ words of high-speed core for the initial $\underline{R}$ arrays and $d_3 d_4 n_{12}$ words for the $\underline{Y}$ arrays. The intermediate random file contains $\overline{d}_1 \overline{d}_2 d_3 d_4 / n_{12} n_{34}$ blocks of size n_{12} by n_{34} which is written and read only once. Maximum efficiency usually requires making the product $n_{12} n_{34}$ as large as possible. Because this integral transformation step involved d^5 operations to transform d^4 integrals it has gained a reputation as a bottleneck in calculations. Actually, however, until d is about 60 the formation of d^4 integrals (over contracted gaussian orbitals) takes longer than the trans-

formation. For larger values of d it is likely that a
CI matrix of large dimension will be formed using
these integrals (or a third or higher order perturbation
calculation will be done). Usually these uses of the
integrals are more time consuming than their production
so the transformation is seldom the limiting step.

   Eigenvalue algorithms. Matrix eigenvalue problems
arise in quantum chemistry at both the SCF and CI level.
The Roothaan SCF method requires solving a non-orthogonal eigenvalue problem of the dimension of the basis
set on each iteration for many of the eigenvalues and
eigenvectors. The CI method usually requires finding
the lowest few eigenvalues of a large matrix in an
orthonormal basis of configurations.
   Several algorithms exist which are suitable for
finding all of the eigenvalues of any matrix of dimension d which can be kept in central memory. The Jacobi
plane rotation method is by far the simplest to program
and is reasonably efficient (34). As it is an iterative
method the running time cannot be rigorously defined,
but times proportional to d^3 are expected. Other
methods usually begin with a non-iterative transformation to tridiagonal form followed by calculation of the
eigenvalues and eigenvectors and a back transformation
to the original problem (34,35). The time required for
the transformations is proportional to d^3 while the time
required to solve the tridiagonal problem is only proportional to d^2.
   The Jacobi method is generally slower than these
other methods unless the matrix is nearly diagonal. In
SCF calculations one is faced with the non-orthogonal
eigenvalue equation

$$\underline{F}\ \underline{C} = \underline{S}\ \underline{C}\ \Lambda$$

where Λ is the diagonal matrix of eigenvalues and $\underline{C}$ is a
matrix of eigenvectors. If an orthogonalizing transfor-
mation W is known such that $W^T S W = 1$, then

$$\underline{W}^T\underline{F}\ \underline{W}\ \underline{W}^{-1}\ \underline{C} = \underline{W}^T\ \underline{S}\ \underline{W}\ \underline{W}^{-1}\ \underline{C}\ \Lambda$$

or

$$\underline{F}'\ \underline{C}' = \underline{C}'\ \Lambda$$

where

$$\underline{F}' = \underline{W}^T\ \underline{F}\ \underline{W}$$

and

$$\underline{C} = \underline{W}\ \underline{C}'$$

Usually on the first iteration of an SCF calculation $\underline{W}$
is computed by the Schmidt orthogonalization method but
thereafter $\underline{W}$ is chosen to be the $\underline{C}$ matrix from the previous iteration. This produces an $\underline{F}'$ matrix which is
nearly diagonal so the Jacobi method becomes quite
efficient after the first iteration. Further, in the
Jacobi method, F' is diagonalized by an iterative
sequence of simple plane-rotation transformations
$\underline{F}'_{(n+1)} = \underline{X}^T_{(n)}\underline{F}'_{(n)}\underline{X}_{(n)}$. The final eigenvectors of F
can thus be generated as $\underline{C} = (\cdots(\underline{W}\underline{X}_{(1)})\underline{X}_{(2)}\cdots\underline{X}_{(n)})$
which avoids the multiplication of $\underline{W}$ by $\underline{C}'$.

A disadvantage of the Jacobi method is that the
error in the eigenvector is usually proportional to
the square root of the error in the eigenvalues. Thus,
in 8 digit arithmetic, only 4 figures can be obtained
in the eigenvectors. The inverse iteration method of
Wilkinson ($\underline{34}$) is a method which gives full accuracy in
the vectors. This method is based on computing the
eigenvector as $(\lambda\underline{1}-\underline{F}'')\underline{C} = \underline{X}$ where λ is the eigenvalue
and $\underline{X}$ is a guess to the eigenvector. Because this
method requires solving a different set of linear
equations for each eigenvector it is only feasible if $\underline{F}''$
has an easily inverted form (solving linear equations
is a d^3 process unless the coefficient matrix has some
simplifying feature). If $\underline{F}''$ is tridiagonal, then the
time for each vector is proportional to d so the time
for d vectors is proportional to d^2.

In CI calculations it is necessary to find a few
solutions to the matrix eigenvalue problem

$$\underline{H}\ \underline{C} = \lambda\ \underline{C}$$

where $\underline{H}$ is of dimension from 10^1 to 10^5. For smaller
dimensions it is most efficient to use the standard
tridiagonalization routines. For matrices which are too
large to fit into high-speed core, special methods have
been developed whose time per eigenvalue is proportional
only to the number of non-zero matrix elements (d^2 at
most). These methods should be useful in other areas of
chemistry as well.

The first development in this area was the Nesbet
method ($\underline{36}$) for finding the lowest (or highest) eigenvalue. This method was reorganized into a better
algorithm by Shavitt ($\underline{37}$) and then extended by Shavitt,
et al. ($\underline{38}$) to find a few non-degenerate eigenvalues.
Recently Davidson ($\underline{39}$) has combined the fundamental
ideas from Nesbet, Lanczos and inverse iteration schemes
to form a method which works for the first few eigenvalues even if they are degenerate. His method, however,

involves a little more input-output than the Nesbet or Shavitt methods.

The basic concept of the Nesbet-Shavitt method is based on iterative sequential optimization of the eigenvector elements. If the quantity $\rho(\underline{C})=\underline{C}^T\underline{H}\underline{C}/\underline{C}^T\underline{C}$ is known for some $\underline{C}^\circ$ and $\rho^\circ=\rho(\underline{C}^\circ)$ is below all the diagonal elements of $\underline{H}$, then sequential minimization of $\rho(\underline{C})$ with respect to each element C_i [i.e. solving $(\partial\rho/\partial C_i)_{C_j^\circ \ldots}=0$ and then stepping C_i° to the new C_i before going to the next value of i] gives

$$\delta C_i = C_i - C_i^\circ = -[(\underline{H}-\rho\underline{1})\underline{C}^\circ]_i/(H_{ii}-\rho)$$

where

$$\rho = \rho(\underline{C}^\circ + \underline{e}_i\delta C_i)$$

while for any value of δC_i

$$\rho(\underline{C}^\circ+\underline{e}_i\delta C_i)-\rho(\underline{C}^\circ) = \frac{2\delta C_i[(\underline{H}-\rho^\circ\underline{1})\underline{C}^\circ]_i + (\delta C_i)^2[H_{ii}-\rho^\circ]}{\underline{C}^{\circ T}\underline{C}^\circ + (\delta C_i)^2 + 2C_i^\circ\delta C_i}$$

Nesbet approximated the optimum δC_i by

$$\delta C_i = -q_i^\circ/(H_{ii}-\rho^\circ)$$

where

$$q_i^\circ = [(\underline{H}-\rho^\circ\underline{1})\underline{C}^\circ]_i$$

while Shavitt found δC_i from the slightly more exact formula

$$\delta C_i = -2q_i^\circ/\{H_{ii}-\rho^\circ+\sqrt{(H_{ii}-\rho^\circ)^2-4q_i^\circ[-q_i^\circ+C_i^\circ(H_{ii}-\rho^\circ)]/\underline{C}^{\circ T}\underline{C}^\circ}\}.$$

Both of these formulas can be shown to give monotonic convergence for ρ. More importantly, Shavitt showed how use of the hermitian property of $\underline{H}$ could be used to write $\underline{H}\underline{C}$ as

$$(\underline{H}\underline{C})_i = \sum_j H_{ij}C_j = \sum_{j\le i} H_{ji}C_i + \sum_{j>i} H_{ij}C_j$$

so that H_{ij} and H_{ji} did not both need to be stored and read from external store. Shavitt et al. further

modified the Nesbet-Shavitt scheme to do excited states
by introducing root-shifting and over-relaxation to
speed convergences. Their method, however, often fails
to converge for nearly degenerate eigenvalues.

Davidson introduced a different method for higher
eigenvalues which also avoids the need to have the elements of $\underline{H}$ stored in any particular order. In this
method the kth eigenvector of $\underline{H}$ for the nth iteration is
expanded in a sequence of orthonormal vectors $\underline{b}_i$,
$i=1\cdots n$ with coefficients found as the kth eigenvector
of the small matrix H with elements $b_i^T H b_j$. Convergence
can be obtained for a reasonably small value of n if the
expansion vectors $\underline{b}$ are chosen appropriately. Davidson
defined

$$\underline{C}_k^{(n)} = \sum_{i=1}^{n} c_{ik}^{(n)} \underline{b}_i$$

$$\underline{q}^{(n)} = [\underline{H}-\rho(C_k^{(n)})\underline{1}]C_k^{(n)}$$

$$\xi_i^{(n)} = q_i^{(n)}/(H_{ii}-\rho)$$

and chose $\underline{b}_{n+1}$ as the normalized residual when $\underline{\xi}^{(n)}$ was
orthogonalized to the preceeding $\underline{b}_1\cdots\underline{b}_n$. This choice
for $\underline{b}_{n+1}$ is similar to the Nesbet choice (and also to
first order perturbation theory and the inverse iteration method). By the excited state variation theorem,
the kth eigenvalue of H as it is sequentially bordered
will decrease monotonically to the kth eigenvalue of $\underline{H}$.
Butscher and Kammer (40) have shown how a slight
modification of this scheme which tracks on certain
large elements of $\underline{C}$ rather than the index k can find a
$\underline{C}$ with a certain desired pattern of coefficients without
prior knowledge of the value of k and without finding
any other eigenvectors.

Literature Cited

1. Hylleraas, E.A., Z. Physik. (1930) 65, 209.
2. James, H.M. and Coolidge, A.S., J. Chem. Phys.
    (1933) 1, 825.
3. Hartree, D.R., Proc. Cambridge Phil. Soc. (1928)
    24, 89.
4. Thomas, L.H. Proc. Cambridge Phil. Soc. (1927)
    23, 542.
5. Slater, J.C., Phys. Rev. (1931) 38, 1109.
6. Pauling, L., J. Am. Chem. Soc. (1931) 53, 1367.

7.   Mulliken, R.S., Phys. Rev. (1928) 32, 186.
8.   Walsh, A.D., J. Chem. Soc. (1953) 2260.
9.   Pauling, L. and Wheland, G.W., J. Chem. Phys. (1933) 1, 362.
10. Pariser, R. and Parr, R.G., J. Chem. Phys. (1953) 21, 466.
11. See for example Karo, A.M. and Allen, L.C., J. Chem. Phys. (1959) 31, 968.
12. Johnson, K.H., Adv. Quantum Chem. (1973) 7, 143.
13. See for example Kouri, D.J., "Energy Structure and Reactivity", Smith, D.W. and McRae, W.B., Eds., John Wiley & Sons, 1973.
14. Boys, S.F., Proc. Roy. Soc. (1950) A200, 542.
15. Hehre, W.J., Stewart, R.F. and Pople, J.A., J. Chem. Phys. (1969) 51, 2657.
16. Dunning, T.H., J. Chem. Phys. (1970) 53, 2823.
17. Shavitt, I., in "Methods in Computational Physics", Vol.2, Alder, B., Fernbach, S. and Rotenberg, M., Academic Press, 1963.
18. Huzinaga, S., Supp. Prog. Theoretical Phys. (1967) 52.
19. Shipman, L.L., and Christoffersen, R.E., Comp. Phys. Comm. (1971) 2, 201.
20. Elbert, S.T., and Davidson, E.R., J. Comput. Phys. (1974) 16, 391.
21. Whitten, J.L., J. Chem. Phys. (1963) 39, 349.
22. Rüdenberg, K., J. Chem. Phys. (1951) 19, 1459.
23. Corbato, F.J., J. Chem. Phys. (1956) 24, 452.
24. Roothaan, C.C.J., Rev. Mod. Phys. (1951) 23, 161.
25. Roothaan, C.C.J. and Bagus, P.S., in "Methods in Computational Physics", Vol. 2, Alder, B., Fernbach, S., and Rotenberg, M., Eds., Academic Press, 1963.
26. Guest, M.F., and Saunders, V.R., Mol. Phys. (1974) 28, 219.
27. Hsu, H., Davidson, E.R. and Pitzer, R.M., J. Chem. Phys., J. Chem. Phys. (1976) 65, 609.
28. For a thorough review see Shavitt, I., in "Modern Theoretical Chemistry", Vol. 2, Schaeffer, H.F. III, Ed., Plenum Press, New York, 1976.
29. Møller, Chr., and Plesset, M.S., Phys. Rev. (1934) 44, 618.
30. Elbert, S.T., and Davidson, E.R., Int. J. Quant. Chem. (1974) 8, 857.
31. Davidson, E.R., "Reduced Density Matrices in Quantum Chemistry, Academic Press, New York, 1976.
32. Davidson, E.R., Int. J. Quant. Chem. (1974) 8, 83.
33. Elbert, S.T., Ab initio Calculations in Urea, Ph.D. thesis, University of Washington, 1973.

34. Wilkinson, J.H., "The Algebraic Eigenvalue Problem",
     Clarendon Press, Oxford, 1965.
35. Givens, J.W., J. Assoc. Comp. Mach. (1957) 4, 298.
36. Nesbet, R.H., J. Chem. Phys. (1965) 43, 311.
37. Shavitt, I, J. Comput. Phys. (1970) 6, 124.
38. Shavitt, I., Bender, C.F., Pipano, A. and Hosteny,
     R.P., J. Comput. Phys. (1973) 11, 90.
39. Davidson, E.R., J. Comput. Phys. (1975) 17, 87.
40. Butscher, W. and Kammer, W.E., J. Comput. Phys.
     (1976) 20, 313.

3

Rational Selection of Algorithms for Molecular Scattering Calculations

ROY G. GORDON

Harvard University, Cambridge, MA 02138

Scattering theory is the link between intermolecular forces, and the various experiments with molecular beams, gases, etc., which depend on collisions between molecules. This link is used in both directions: In the theoretical approach the intermolecular forces are used to predict the outcome of experiments. In the empirical approach, experimental results are inverted or analyzed to obtain information about the intermolecular potential.

For most molecular scattering phenomena, it is usually assumed that nonrelativistic quantum mechanics provides an accurate description. Therefore, one might expect the field of molecular collision phenomena to be nicely unified by the application of nonrelativistic quantum-mechanical scattering theory. Instead, one finds a bewildering variety of methods, approximations, techniques, formulations and reformulations are used to treat molecular collisions. One might be tempted to blame this multitude of approaches on the conceit of the many theoreticians who have worked in this area, each developing his own point of view. In fact, this variety is more nearly due to the following two circumstances: 1. Exact quantum mechanical scattering calculations are not yet feasible for all types of molecular collisions. Therefore some types of approximations are necessary to treat the quantum mechanically intractable cases. 2. The very richness and variety of molecular scattering processes require that a number of different approximation methods be used in different situations.

We believe that suitable methods have in fact been developed, to treat successfully almost all types of molecular collisions. The question thus arises: How do we select the most appropriate method for a given problem? In Section II we discuss some criteria for choosing between methods. In section III we propose an explicit algorithm for selecting the best available method for a given collision process, and for a given set of experiments measuring that process. Then we apply this algorithm to a number of examples, mainly from inelastic scattering. It is hoped that these examples will illustrate the way in which one should choose between methods, and the kind of information such a choice requires.

In addition, the examples described in Section III are all chosen to represent real cases for which calculations have been completed, or are in progress. Thus, they provide a guide to some recent applications of each of the methods discussed, and the reader himself can evaluate the state of the art in applications of each method.

Criteria for Choosing an Appropriate Scattering Theory

In order to make a rational selection of a scattering theory to apply to a specific problem, we must formulate criteria upon which this choice is to be based. It seems to us that there are three main considerations:

Feasibility. It is necessary that the method be applicable, in a practical sense, to the problem of interest. Difficulties may occur at various stages: analytic difficulties (e.g. in evaluating matrix elements, or in transforming coordinate systems); exceeding memory size or running time of computers; difficulties in averaging and analysis of results into a form to compare with experiments.

Accuracy. The results must be sufficiently accurate to interpret the experiments of interest. In a complete quantum-mechanical calculation, this accuracy can be verified by convergence tests within the calculation. In classical, or other approximate methods, accuracy and reliability generally must be judged by experience with test comparisons with complete quantum-mechanical calculations. The numerical stability of the method must also be considered.

Ease of Calculation. When more than one method meets the above criteria of feasibility and accuracy, one has the luxury of choosing the easiest of the possible methods. Some considerations in the "case" of calculation might include the following: If the evaluation of the interaction potential is difficult (as it is likely to be in any realistic case), one would prefer the method which requires the smallest number of values of the potential. Other considerations might be the complexity and cost of the computer calculations, and the availability of well-documented and reliable computer programs.

Next we must discuss the specific methods of calculation which we shall recommend, in the light of the three criteria discussed above.

Quantum Scattering ("close coupling") (1). The feasibility of a full quantum scattering calculation depends mostly upon the

number (N_c) of internal states which are coupled together by the
interaction potential, during the strongest part of the collision.
The most efficient quantum scattering method currently available
is based on piecewise analytic solution to model potentials which
approximate the true potential to any prescribed degree of
accuracy (2). Piecewise linear model potentials usually provide
sufficient accuracy, along with an accurate and efficient algorithm for the calculations (2). More accurate model potentials
can now be based on piecewise quadratic approximations, for which
an effective solution algorithm has now been devised (3). While
one can program this method to work with whatever size computer
is available (using disk storage if necessary), the number of disk
accesses becomes rather large unless the computer memory is large
enough to store at least eight N_c by N_c matrices (8 N_c^2 numbers).
Up to about 100 N_c^3 multiplications and additions are required to
construct a single scattering matrix. These storage and timing
restrictions currently restrict feasible calculations to N_c about
100 or less. Thus a number of approximations are being explored,
which may reduce the number N_c. These include the use of effective Hamiltonians (4-9) and j_z conserving approximations (10-12).
Very promising results are being obtained, and these approximations should allow the use of quantum scattering methods to be
used for a much wider range of molecules.

The accuracy of the quantum scattering results is limited
mainly by the number of internal states included (close-coupling
approximation). Therefore one must check that the predictions of
interest converge as one increases the number of internal states.
The accuracy of the radial integration can be set at any pre-
determined value. Further work (13) has simplified the perturbation formulas for setting the accuracy of the radial integration.
The method was constructed to be numerically stable, and in
practice not more than two digits are lost in roundoff error, even
in calculations involving millions of arithmetic operations.

As for ease of calculation, only a small number (say 30) of
radial integration points are required, so that not too many
evaluations of the potential are necessary. A complete computer
program for quantum-mechanical elastic and inelastic scattering is
available (14).

The quantum theory of reactive scattering is not as highly
developed as for inelastic scattering. No generally applicable
algorithm has yet been perfected, particularly for three-dimensional reactions. However, many promising approaches are being
explored.

Distorted Wave Born Approximation. Quantum scattering calculations are sometimes made using the distorted wave Born approximation (15). Such calculations have the advantage of almost
always being feasible numerically. For simple cases, one can also
obtain some results analytically (16). However, the accuracy of
the results is generally poor, for most molecular collisions. A

necessary condition for the results to be accurate, is that all
the calculated transition probabilities be small compared to
unity. However, this is not a sufficient condition, since small
transition probabilities can result from fortuitous cancellation
of large negative and positive contributions to the perburbation
integrals. One can test for this possibility by checking whether
the sum of all the perturbation integrals remains small as we
build them up by adding on contributions from the various radial
intervals. This provides both a necessary and sufficient condition for the validity of perturbation theory.

Classical Mechanics. The description of scattering by
classical mechanics has the important advantage of almost always
being feasible to carry out. Only three circumstances occasionally make it difficult to obtain results with classical scattering
theory: 1) There may be points at which the coordinates chosen
for integration become singular or undefined (17). If a trajectory approaches one of these points, the numerical integration may
break down. Such difficulties may be avoided by changing coordinate systems. 2) If some coordinates change much more rapidly
than others, the equations become difficult to integrate numerically. These difficulties may be reduced by using action-angle
coordinates for the rapidly varying coordinates (18), and by using
a very stable and accurate integration technique, such as Runge-
Kutta. 3) Some trajectories in both inelastic (19) and
reactive (20) collisions are long and complicated, corresponding
to resonances or long-lived collision complexes. Unless one
really needs to know the details of such collisions, it is probably best to use a statistical theory to describe the distribution of results for these collisions.

Semiclassical Methods. The accuracy of classical calculations is usually adequate when the experiments of interest average
over at least several quantum states. If, however, no classical
trajectories connect the initial and final states of motion, the
classical prediction is a vanishing cross section or rate constant
for that process. The correct quantum-mechanical prediction may,
however, be a small but non-zero rate for such a "classically
forbidden" process. "Tunneling" through a potential barrier is a
simple example. The connection formulas in the WKB method may be
viewed as providing a complex-valued trajectory which does link
the "classically forbidden" states. In the WKB treatment, the
probability for passing through this complex trajectory, is
related to the exponential of the imaginary part of the classical
action function accumulated along the complex path. Recently,
this treatment has been generalized to inelastic and reactive
scattering (21-24). The main difficulty at present in applying
this method, is finding the actual complex trajectories in a
numerically stable way. Several approaches have been suggested,
and this is an active field of current research. One should note

that the method appears also to require that the interaction
potential be an analytic function of all its coordinates, so that
it, too, can be analytically continued. Whether a continuation
method can be applied to a potential defined by a table of numerical values and some interpolation formulae, is not clear at
present. Another practical problem with the semiclassical
method, is the numerical search for trajectories with specific
(quantized) values of the initial and final momenta (quantum
numbers). For molecules with several internal degrees of freedom,
this may be a difficult task. Furthermore, if there are more than
several trajectories with the same initial and final quantum
numbers (as is typically the case when the trajectories are
complicated), then the semiclassical results may not be very
accurate.

When classical mechanics is applied to experiments involving
only one or two quantum states, the results are generally less
accurate than for the cases involving averages over many quantum
states. However, even simple correspondence principle aruguments,
assigning classical results to the quantum state of nearest
angular momentum, predict line-broadening cross sections to an
accuracy comparable to the experimental uncertainty (19,25-27).
Moreover, by including interference effects between different
trajectories (28-32), one can make fairly accurate predictions for
elastic (28) vibrationally (33) and rotationally (34) inelastic,
and reactive (35) scattering. This is a very useful approach,
which will certainly be used more in future calculations, to
improve the accuracy of classical predictions. The semiclassical
approach has been reviewed recently by Connor (36).

Classical Path. Another approach to scattering calculations
uses a quantum-mechanical description of the internal states, but
classical mechanics for the translational motion. This "classical
path" method has been popular in line-shape calculations (37,38).
It is almost always feasible to carry out such calculations in the
perturbation approximation for the internal states (37). Only
recently have practical methods been developed to perform non-
perturbative calculations in this approach (39).

To get accurate results from this approach, it is necessary
that the collisional changes in the internal energy be small
compared to the translational energy. Then one can accurately
assume a common translation path for all coupled internal states.
In the usual applications of this method, one does not include
interference effects between different classical paths, so that
translational quantum effects, including total elastic cross
sections, are not predicted. If the perturbation approximation is
also used, accuracy can be guaranteed only when the sum of the
transition probabilities remains small throughout the collision.

These classical path calculations are relatively easy to
carry out, and analytic results are available in the straight-line
path, perturbation limit (40). Thus when the approximations are

valid, this classical path approach should be used.

An Algorithm for Choosing an Appropriate Scattering Theory

Using the criteria discussed above, we wish to select the easiest method of calculation which is both feasible to apply to the molecules of interest, and whose results are sufficiently accurate to describe the relevant experimental results. We have found it convenient to organize this selection process into a flow chart, which is given in Fig. 1. Starting at the top, one makes a sequence of decisions based upon the criteria for feasibility and accuracy. Decisions about the relative ease of different methods are not made explicitly; they are implicit in the organization of the flow chart.

When one's path in the flow chart reaches a box with no lines going out from it, and double underlines at its bottom, one has arrived at the most suitable method. In some cases, one's decision at some point may be conditional on a variable in the problem. For example, transition probabilities may be small compared to unity for large orbital angular momenta, but not for small ones. In such cases one should follow both branches of the decision, and arrive at two different methods, one for each range of the variable. In a few such cases, both branches may later rejoin, and only one method is recommended after all. In more difficult cases, as many as three different methods have been found to be necessary for different ranges of the variables. Examples of all these cases have been found.

We first follow the flow chart for the simple case of elastic scattering of structureless atoms. The number of internal states, N_c, is one, quantum scattering calculations are feasible and recommended, for even the smallest modern computer. The Numerov method has often been used for such calculations [41], but the recent method based on analytic approximations by Airy functions [2] obtains the same results with many fewer evaluations of the potential function. The WKB approximation also requires a relatively small number of function evaluations, but its accuracy is limited, whereas the piecewise analytic method [2] can obtain results to any preset, desired accuracy.

Next we consider rotationally inelastic scattering of H_2 with He. At room temperature, the maximum rotational angular momentum state which is significantly populated is $j_{max} = 4$. Thus we estimate $N_c = (j_{max}/2 + 1)^2 = 9$, including all the m-states. The data storage $8N_c^2$ is less than 1000 numbers, only a small addition to the quantum scattering program code (about 100 K-bytes). Assuming a multiply time of 1 μ-sec., $100 N_c^3$ is less than 0.1 sec computer time per $\underline{S}$ matrix. Thus the quantum scattering calculations are quite practical, and have been carried out for more than a dozen different potential surfaces [42]. The results are in good agreement with molecular beam results, sound absorption, and line shapes in light scattering and NMR. Because of the wide

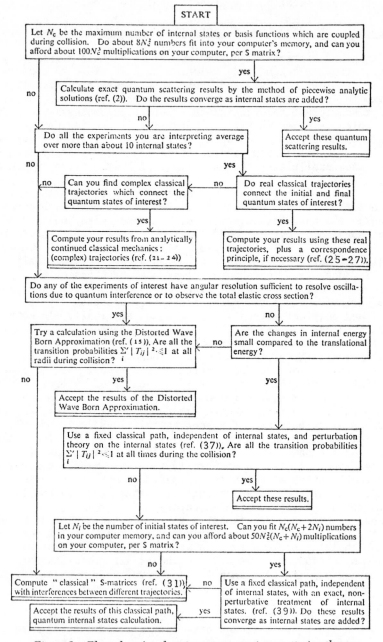

Figure 1.   Flow chart for choosing an appropriate scattering theory

spacing of the rotational levels, and the relatively weak angle-dependent potential, these results converge very quickly as j_{max} increases, and $j_{max} = 4$ is adequate for all the experiments at temperatures up to 300°K.

For collisions of H_2 with atoms at higher energies, both vibrational and rotational excitation occurs. At 1 eV, about 50 channels are open. For a complete quantum scattering calculation, we estimate data storage at $8N_c^2 \approx 20,000$ single precision words, and computer time of 12 sec per $\underline{S}$ matrix (again assuming a 1 μ-sec multiply time). Convergence is obtained with the addition of a few closed channels, and such calculations are feasible, and have recently been carried out for H_2 + He (43), and H_2 + Li^+ (44).

For vibrational and rotational relaxation of D_2 at 1 eV, about 140 channels are open, so the quantum scattering estimates are about 160,000 numbers in data storage, and about 5 min computing time per $\underline{S}$ matrix, or 2 sec per initial condition. While such calculations are feasible on a large computer, they might be too expensive. Then, if one is averaging over rotational states to find vibrational transition probabilities, the flow chart suggests classical trajectories. However, the vibrational coupling is so weak that no real trajectories connect different vibrational states, so complex trajectories must be calculated to find the vibrational transition probabilities (45). One should note, however, that if one wants to find all the individual rotation-vibration transition probabilities, the quantum calculation, at 2 sec per initial condition, uses less computer time than the complex trajectory calculation, which requires about 2 sec per complex trajectory, and a search of several complex trajectories for each initial condition.

If we consider the collisions of two molecules (rather than atom + molecule, as above), the number of coupled channels is approximately the _square_ of the number of accessible internal states of either molecule separately. Thus for rotational excitation of two hydrogen molecules near room temperature, $N_c \underset{\sim}{\sim}$ $(j_{max}/2 + 1)^4 = 81$ for $j_{max} = 4$, and quantum calculations are feasible. However, for vibration-rotation transitions at 1 eV, 50 internal states for each molecule correspond to $N_c = 2500$ channels, and exact quantum calculations are not feasible. If we want individual transition probabilities for this case, the flow chart brings us to try the distorted wave Born approximation, which is feasible and accurate for this case.

Next we consider some more difficult cases, in which several methods are recommended for different parts of the calculation. For rotational excitation of HCl by Ar at room temperature, the maximum rotational angular momentum quantum number coupled during collision is about 12. The maximum number of coupled j,m states is $N_c = (j_{max} + 1)(j_{max} + 2)/2 = 91$, since HCl is a heterodiatomic molecule, and thus all states of the same total parity are coupled. With 91 channels, the quantum scattering calculations are feasible, but rather expensive. A further complication of the

quantum calculations for this case, is the fact that many bound
states of HCl + Ar exist, which will lead to many resonances ·in
the scattering, and thus difficult energy averaging the cross
sections. Thus we explore the alternative methods with the flow
chart. For interpreting infrared line-widths, we average over the
2j + 1 m-states. For an initial j greater than 5 we thus average
over enough m states so that the classical method, plus the
correspondence principle, is adequate for these cases. For the
low-j lines, we observe that in the absence of differential cross
section measurements, we do not require a "high resolution"
quantum calculation. The rotational energy changes, for the low
j states, are small compared to the typical translational energies, so the fixed classical path approximation is valid. For
collisions at large impact parameter, the classical path-pertur-
bation theory results are of acceptable accuracy. However, for
small impact parameter cases, the perturbation theory fails. To
select a method for the remaining cases we note that the maximum
number of coupled initial states up to j = 5 is N_c = (j + 1)
(j + 2)/2 = 21. The storage estimates for a non-perturbative
classical path calculation are thus 91(91 + 2x21) $\approx$ 21,000
numbers, and computer time 50(91)^2(91 + 21) x 10^{-6} sec = 46 sec
per $\underline{S}$ matrix. This classical path method is thus feasible for the
remaining initial conditions, and has been used (39) to calculate
infrared and NMR line shapes for this system.

For a heavier system, such as N_2O + Ar, a calculation of
rotational transitions and microwave or infrared line widths would
follow the same course through the flow chart, as that followed
above in detail for HCl + Ar. However, at the last stage (low j,
small b collisions), the number of coupled states would probably
be too large for the non-perturbative, fixed classical path
calculation to be practical. Then one should calculate "classical
$\underline{S}$ matrices" including interference between trajectories, to cover
these remaining collisions.

Conclusion

The theory of molecular scattering has now been developed to
the point that scattering calculations can be made with an
accuracy sufficient for comparison with current experiments. Thus
any discrepancy between theory and experiment should be traced to
an inadequate knowledge of the interaction potentials, or to
experimental errors, rather than to approximations in the
collision dynamics. This tighter coupling of theory and experiment should permit a much more fruitful utilization of the results
of molecular beam scattering.

Abstract

A critical discussion is given of some of the more useful and
accurate methods for the calculation of cross sections for various

types of molecular collisions. Quantum mechanical, classical and semiclassical methods are considered. Criteria are summarized for the feasibility of various calculations, and for the accuracy of the results. A flow chart is formulated, which uses these criteria to select, for given molecules and types of experiments, the easiest calculational algorithm which yields accurate results. Examples of this selection process are given, drawn mainly from recent calculations of inelastic scattering.

Acknowledgments

This work was supported in part by the National Science Foundation. It is based on a paper delivered at the General Discussion on Molecular Beam Scattering, 16th - 18th of April, 1973, with some additional comments and more recent references.

Literature Cited

1. For reviews of recent quantum scattering methods, see "Methods in Computational Physics," B. Alder et al., Ed., Vol. 10, Academic Press, New York, 1971.
2. Gordon, R.G., ibid., Chap. 2, p. 81.
3. Luthey, Z., thesis (Harvard University, 1974).
4. Rabitz, H., J. Chem. Phys. (1972) 57, 1718.
5. Zarur, G. and Rabitz, H., J. Chem. Phys. (1973) 59, 943.
6. ibid. (1974) 60, 2057.
7. Englot, G. and Rabitz, H., Chem. Phys. (1974) 4, 458.
8. Englot, G. and Rabitz, H., Phys. Rev. (1974) A10, 2187.
9. Reviewed by H. Rabitz, "Modern Theoretical Chemistry III," W.H. Miller, Ed. (to appear, 1976).
10. McGuire, P., Chem. Phys. Lett. (1973) 23, 575.
11. Kouri, D.J. and McGuire P., Chem. Phys. Lett. (1974) 29, 414.
12. McGuire, P. and Kouri, D.J., J. Chem. Phys. (1974) 60, 2488.
13. Rosenthal, A. and Gordon, R.G., J. Chem. Phys. (1976) 64, 1621.
14. Program No. 187, Quantum Chemistry Program Exchange, Chemistry Department, Indiana University, Bloomington, Indiana 47401, U.S.A.
15. See, for example, Rodberg, L.S. and Thaler, R.M., "Introduction to the Quantum Theory of Scattering," Chap. 12, Academic Press, New York, 1967.
16. Starkschall, G. and Gordon, R.B., to be published.
17. Cross, R.J., Jr., and Herschbach, D.R., J. Chem. Phys. (1965) 43, 3530.
18. Cohen, A.O. and Gordon, R.G., to be published.
19. Pearson, R. and Gordon, R.G., to be published.
20. Brumer, P.W. and Karplus, M., to be published.
21. Miller, W.H. and George, T.F., J. Chem. Phys. (1972) 56, 5668.
22. ibid., 5722.

23. Doll, J.D. and Miller, W.H., J. Chem. Phys. (1972) 57, 5019.
24. Marcus, R., Kreek, H.R. and Stine, J.R., Disc. Faraday Soc. (1972).
25. Gordon, R.G., J. Chem. Phys. (1966) 44, 3083.
26. Gordon, R.G. and McGinnis, R.P., ibid. (1971) 55, 4898.
27. Bunker, D.I., ref. 1, Chap. 7.
28. Ford, K.W. and Wheeler, J.A., Ann. Phys. (N.Y.) (1959) 7, 259.
29. Miller, W.H., J. Chem. Phys. (1970) 53, 1949.
30. ibid., 3578.
31. Miller, W.H., Adv. Chem. Phys. (1974) 25, 69.
32. Marcus, R.A., J. Chem. Phys. (1972) 57, 4903, and references therein.
33. Miller, W.H., Chem. Phys. Lett. (1970) 7, 431.
34. Miller, W.H., J. Chem. Phys. (1971) 54, 5386.
35. Rankin, C.C. and Miller, W.H., J. Chem. Phys. (1971) 55, 3150.
36. Connor, J.N.L., Chemical Society Reviews 5, 125.
37. Anderson, P.W., Phys. Rev. (1949) 76, 647.
38. For a recent review see Birnbaum, G., Adv. Chem. Phys., (1967) 12, 487.
39. Neilsen, W. and Gordon, R.G., J. Chem. Phys. (1973) 58, 4131.
40. Cross, R.J. and Gordon, R.G., J. Chem. Phys. (1966) 45, 3571.
41. Cooley, J.W., Math. Computation (1961) 15, 363.
42. Shafer, R. and Gordon, R.G., J. Chem. Phys. (1973) 58, 5422.
43. Eastes, W. and Secrest, D., J. Chem. Phys. (1972) 56, 640.
44. van den Bergh, H., David, R., Fraubel, M., Fremerey, H. and Toennies, J.P., Disc. Faraday Soc. (1972).
45. Miller, W.H., Disc. Faraday Soc. (1972).

Molecular Dynamics and Transition State Theory: The Simulation of Infrequent Events

CHARLES H. BENNETT

IBM Thomas J. Watson Research Center, Yorktown Heights, NY 10598

Before the advent of the high speed digital computer, the theoretical treatment of atomic motion was limited to systems whose dynamics admitted an approximate separation of the many-body problem into analytically tractable one- or two-body problems. Two approximations were the most useful in making this separation:

1) Stochastic approximations such as 'random walk' or 'molecular chaos', which treat the motion as a succession of simple one- or two-body events, neglecting the correlations between these events implied by the overall deterministic dynamics. The analytical theory of gases, for example, is based on the molecular chaos assumption, i.e. the neglect of correlations betweeen consecutive collision partners of the same molecule. Another example is the random walk theory of diffusion in solids, which neglects the dynamical correlations between consecutive jumps of a diffusing lattice vacancy or interstitial.

2) the harmonic approximation, which treats atomic vibrations as a superposition of independent normal modes. This has been most successfully applied to solids and free molecules at low temperatures, where the amplitude of oscillation is small enough to remain in the neighborhood of a quadratic minimum of the potential energy.

Transition state theory (1), the traditional way of calculating the frequency of infrequent dynamical events (transitions) involving a bottleneck or saddle point, typically had to call on both these approximations before yielding quantitative predictions.

Because of the unavailability of a method for solving the classical many-body problem directly, the harmonic approximation was sometimes stretched, or stochastic behavior assumed too early, in an effort to

predict equilibrium thermodynamic properties, transport
coefficients, and transition rates in systems that were
too strongly coupled and too anharmonic for the results
to be reliable. In the last two decades this situation
has been radically changed by the ability of computers
to integrate the classical equations of motion (the
classical trajectory or molecular dynamics 'MD' technique, cf refs. 2, 3, 4, and reviews 5, 6) for systems
of up to several thousand particles, thereby making it
possible to attack by direct simulation such previous-
ly-intractable problems as the equilibrium and transport properties of liquids and hot anharmonic solids,
chemical reactions in gases, the structure of small
droplets, and conformational rearrangements in large
molecules. In addition to providing dynamical information, the molecular dynamics method (as well as the
Monte Carlo (MC) method of Metropolis et. al. 7, 8, 6)
is routinely used to calculate equilibrium thermodynamic properties in many of the same systems (especially
liquids), when these cannot be obtained anaytically.
The scope of these classical simulation techniques is
determined by a number of considerations:

1) They are not applicable to strongly quantum mechanical systems, like liquid or solid H or He, in
which the thermal de Broglie wavelength ($h/\sqrt{2\pi mkT}$) is
comparable to the atomic dimensions;

2) They are unnecessary when the harmonic or random
walk approximations are valid, (e.g. in calculating the
thermodynamic properties of cold solids or dilute gases).

3) The potential energy surface (i.e. the potential
energy expressed as a function of the atomic positions)
on which the classical trajectory moves is almost always semi-empirical and rather imprecisely known, because accurate quantum mechanical claculations of it
are impossibly expensive except in the simplest systems. For use in a MD or MC program, the potential
energy must be rendered into a form (e.g. a sum of
two-body and sometimes three-body forces) that can be
evaluated repeatedly at a cost of not more than a few
seconds computer time per evaluation.

4) The methods are of course restricted to simulating
systems of microscipic size (typically between 3 and
10,000 atoms). This is not a very serious limitation
because on the one hand, with existing algorithms, simulation cost increases only a little faster than lin-
early with the number of atoms; and, on the other hand,
a system of 1000 atoms or less is generally large
enough to reproduce most macroscopic properties of matter, except for long range fluctuations near critical

points.

5) The most serious practical limitation of molecular dynamics comes from its slowness: for a small (10-20 atom) system each second of computer time suffices to simulate about 1 picosecond of physical time, whereas one is often interested in simulating phenomena taking place on a much longer time scale. This problem is not merely a matter of existing computers being too slow-- indeed, 1 to 10 picoseconds per second is about as fast as one can comfortably watch an animated display of molecular motion-- rather it is a manifestation of a common paradox in molecular dynamics: concealment of the desired information by mountains of irrelevant detail.

The bulk of this chapter will expound a synthesis of molecular dynamics (and Monte Carlo) methods with transition state theory that combines the former's freedom from questionable approximations with the latter's ability to predict arbitrarily infrequent events, events that would be prohibitively expensive to simulate directly. However, before beginning this exposition, a few more philosophical remarks will be made on the irony of being able to simulate molecular motion accurately on a picosecond time scale, without thereby being able to understand the consequences of that motion on a 1 second time scale. To exhibit the irony in an extreme form, consider a system whose simulation is somewhat beyond the range of present molecular dynamics technique: a globular protein (e.g. an enzyme) in its normal aqueous environment. An animated movie of this system could not be run much faster than 10 picoseconds per second (10 psec. is approximately the lifetime of a hydrogen bond in water) without having the water molecules move too fast for the eye to follow. At this rate, a typical enzyme-catalyzed reaction would take several years to watch, and the spontaneous folding-up of the globular protein from an extended polypeptide chain would take thousands of years. The calculation necessary to make the movie would of course take several several orders of magnitude longer on present computers; but even if speed of computation were not a problem, watching such a long movie would be.

It is hard to believe that, in order to see how the enzyme works, or how the protein folds up, one must view the movie in its entirety. It is more plausible that there are only a few interesting parts, during which the system passes through critical bottlenecks in its configuration space; the rest of the time being spent exploring large, equilibrated reservoirs between the bottlenecks. If the trajectory calculation were

repeated many times, starting from slightly different
initial conditions, one would expect the trajectory to
pass through the same critical bottlenecks in the same
order, but the less constrained portions of the trajectory, between bottlenecks, would probably be different
each time. An adequate understanding of the relaxation
process as a whole could therefore be gained by gather-
ing dynamical information on trajectories in the neighborhood of each critical bottleneck, and supplementing
this by a statistical characterization (in terms of a
first-order rate constant, or its reciprocal, a mean
residence time) of each intervening reservoir. Before
accepting the hypothesis that only a few parts of the
movie would be interesting enough to call for detailed
dynamical simulation, let us consider the two remaining
possibilities for a thousand-year movie, viz. uniformly dull, and uniformly interesting.

The uniformly dull movie would depict a slow, uni-
formly-progressive relaxation process, like the diffusion of impurities into a homogeneous medium or the
fall of sand through an hourglass. Such a relaxation
process has no single bottleneck (or, equivalently, has
very many small equal bottlenecks), but it is only
likely to occur in a system that possesses some obvious
structural uniformity (in the cases cited, the uniformity of the medium into which diffusion occurs, or the
uniformity of the sand grains), which would account in
a natural way for the uniform rate of progress at different degrees of completion. More precisely and res-
trictively, the uniform slow progress can usually be
measured by one or a few slowly-relaxing,
'hydrodynamic' degrees of freedom, whose equations of
motion can be solved independently of the other degrees
of freedom. In the hourglass example, the mean height
of the sand is such degree of freedom; its approximate
equation of motion can be solved without reference to
the detailed trajectory, which passes through a new
bottleneck in configuration space every time a grain of
sand falls through the bottleneck in real space.

A movie of a such a hydrodynamic relaxation process has no really exciting parts, but all parts are
more or less typical, and an understanding of the process as a whole can be gained by viewing a few parts
(say at the beginning, middle, and end), and interpo-
lating between them by the equations of motion for the
slow degrees of freedom. The detailed sequence of
bottlenecks-- e.g. the order in which the sand grains
fell-- is not reproducible by this procedure, but
neither is it important. The connection between molecular dynamics and hydrodynamics in uniform fluids is

of considerable current interest (9), but it is peripheral to the subject of this review, viz. relatively
fast but infrequent events, particularly those occurring in spatially nonuniform systems, whose lack of
symmetry practically guarantees that a few bottlenecks
will be much harder than all the rest.

Undoubtedly there are systems that suffer both
from bottlenecks and slow modes, e.g. any sizeable
change in a the conformation of a protein involves many
atoms and is damped by the viscosity of the surrounding
water; thus, even in the absence of any activation barrier, it would have a relaxation time several orders of
magnitude longer than that of a single water molecule.
However, really large disparities in time scale, e.g.
10 orders of magnitude in a system of a few thousand
atoms, cannot result from hydrodynamic modes alone, but
must be due chiefly to bottlenecks.

The final possibility, a uniformly interesting
movie, would have to depict a process with thousands or
millions of critical steps occuring in a definite order, each step necessary to understand the next, as in
an industrial process, the functioning of a digital
computer, or the development of an embryo. Enzymes,
having been optimized by natural selection, may be expected to have somewhat complex mechanisms of action,
perhaps with several equally important critical steps,
but not with thousands of them. There is reason to
believe that processes with thousands of reproducible
non-trivial steps usually occur only in systems that
are held away from thermal equilibrium by an external
driving force. They thus belong to the realm of complex behavior in continuously dissipative open systems,
rather than to the realm of relaxation processes in
closed systems.

Transition State Theory and Molecular Dynamics

The idea of characterizing infrequent events in terms
of a bottleneck or saddle point neighborhood is much
older than the digital computer, and indeed is the basis of transition state theory (TST), developed in the
thirties (1) and since then applied to a wide range of
relaxation phenomena ranging from chemical reactions in
gases to diffusion in solids. Unfortunately, before
the feasibility of large scale Monte Carlo and dynamic
calculations, transition state theory could not be developed to the point of yielding quantitative predictions without making certain simplifying assumptions
which usually were not theoretically justified, although they often worked well in practice. Three so-

mewhat related assumptions were generally made:
1) that the bottleneck is an approximately quadratic
portion of the potential energy surface containing a
single saddle point (i.e. a point where the the first
derivative, ∇U, of the potential energy is zero and
where its second derivative matrix, $\nabla\nabla U$, has exactly
one negative eigenvalue). For this (harmonic) approximation to be justified, the nearly-quadratic portion of
the potential energy surface should extend at least kT
above and below the exact saddle point.
2) that the typical trajectory does not reverse its
direction while in the saddle point neighborhood (in
other words, the transmission coefficient is 100 per
cent).
3) that an equilibrium distribution of microstates
prevails in the saddle point neighborhood, even when
the system as a whole is in a non-equilibrium macrostate, with trajectories approaching the saddle point
from one side ('reactant') but not the other
('product').

By marrying molecular dynamics to transition state
theory, these questionable assumptions can be dispensed
with, and one can simulate a relaxation process involving bottlenecks rigorously, assuming only 1) classical
mechanics, and 2) local equilibrium within the reactant and product zones separately. For simplicity we
will first treat a situation in which there is only one
bottleneck, whose location is known. Later, we will
consider processes involving many bottlenecks, and will
discuss computer-assisted heuristic methods for finding
bottlenecks when their locations are not known
a priori.

The essential trick for doing dynamical simula-
tions of infrequent events, discovered by Keck (10), is
to use starting points chosen from an equilibrium distribution in the bottleneck region, and from each of
these starting points to generate a trajectory by integrating Newton's equations both forward and backward
in time; rather than to use starting points in the
reactant region and compute trajectories forward in
time, hoping for them to enter the bottleneck. One
thus avoids wasting a lot of time calculating trajectories that do not enter. Furthermore, although the trajectories are originally calculated on the basis of an
equilibrium distribution in the bottleneck, this distribution can be rigorously corrected, using information provided by the trajectories themselves, to reflect the situation in a bottleneck connecting two reservoirs not at equilibrium with each other.

The system in which the transitions are occuring

will be assumed to be a closed system consisting of
N = several to several thousand atoms, describable by a
classical Hamiltonian

$$H = (\sum_{i=1}^{3N} p_i^2 / 2m_i) + U(q_1, q_2 \ldots q_{3N}), \qquad (1)$$

where q_i denotes the i'th atomic cartesian coordinate, and m_i its mass, and where $U(\underline{q})$ is the potential energy function discussed earlier. The system
will be assumed to have no constants of motion other
than the energy: linear momentum, even if conserved,
affects the dynamics only in a trivial manner; while
angular momentum is not conserved in the presence of
periodic boundary conditions (these are ordinarily used
in molecular dynamics work on condensed systems to
abolish surface effects). It is often convenient to
define the Hamiltonian in terms of mass-weighted
coordinates, $\underline{q} \leftarrow \underline{q}/\sqrt{m}$, so that the equilibrium velocity distribution becomes isotropic, and the dynamics is
simply that of a particle rolling on the potential energy surface: $\ddot{\underline{q}} = -\nabla U(\underline{q})$.

The Question of Equilibrium in the Bottleneck.
This question will be discussed at some length (see
also Anderson, ref. 11), because it has been the source
of much confusion in the past. Consider a closed system whose 6N-dimensional phase space contains two regions arbitrarily labelled 'reactant' and 'product', as
well as a third 'bottleneck' region placed so as to
intersect essentially all trajectories passing between
the other two regions.

Figure 1

	:		:	
A	:	B	:	C
(reactant)	:	(bottleneck)	:	(product)
	:		:	

Since $\underline{A}$, $\underline{B}$, and $\underline{C}$ are regions in the phase space of a
single closed system, the transitions between A and $\underline{C}$
represent a unimolecular reaction or isomerization,
rather than a general reaction in the sense of chemical
kinetics. Unlike some unimolecular reactions, (e.g the
decomposition of diatomic molecules) the molecular dynamics system of eq. 1 will be assumed to have sufficiently many well-coupled degrees of freedom that transitions between reactant and product regions occur
spontaneously, without outside interference.

First let us assume that the system has been un-
disturbed for so long that it is in a macrostate of
thermal equilibrium. Trajectories will then pass
through the bottleneck region equally often from left
to right and from right to left, and the probabilities
of different microstates in the bottleneck region, as
in any part of phase space, will be given by the formulas of equilibrium statistical mechanics (e.g. the
equilibrium microcanonical density,

$$P_{eq}(\underline{p},\underline{q}) = \frac{\delta(H(\underline{p},\underline{q})-E)}{\int d\omega\ \delta(H(\underline{p},\underline{q})-E)}, \tag{2}$$

for a system whose equations of motion conserve energy
but not linear or angular momentum). In the denominator $d\omega$ represents the 6N dimensional volume element
$(dp_1 \cdot dp_2 \cdot \ldots dp_{3N} \cdot dq_1 \cdot dq_2 \cdot \ldots dq_{3N})$.
The equilibrium distribution in the bottleneck
region is a rigorous result for any system in macroscopic equilibrium and does not depend on how easy or
difficult the bottleneck is to enter, or on how quickly
the typical trajectory passes through. Nevertheless,
it has seemed intuitively implausible to some solid
state physicists (12), who have argued that the typical
atom, in making a diffusive jump, usually approaches
the saddle point so quickly that the neighboring atoms
(between which the jumping atom must pass) do not have
time to relax outward fully, as they would have, had
the jumping atom been brought to the saddle point slowly and allowed to equilibrate there. The error here is
in regarding the jumping atom's approach solely as a
cause of the outward relaxation, when it may equally
well be a result of that relaxation, inasmuch as prior
outward relaxation of the neighbors makes it easier for
the jumping atom to pass through. The jump event is
more properly treated as a fluctuation in a many-body
system at thermal equilibrium: the jumping atom's presence in the saddle point neither causes, nor results
from, but rather is instantaneously correlated with, a
relaxation in the mean positions of all other atoms in
the system. Similar arguments imply that the velocity
distribution of atoms found in the saddle point neighborhood is thermal and Maxwellian. Although a jumping
atom will usually need more-than-average kinetic energy
to ascend to the saddle point, all this excess kinetic
energy will, on the average, have been converted into
potential energy during the ascent, only to be recov-
ered as kinetic energy during the descent.

Now consider a nonequilibrium macrostate in which reactant and product zones are not in equilibrium with each other, but each by itself is in equilibrium. Strictly speaking this condition cannot maintain itself if there is any flux through the bottleneck--in the long run global equilibrium will of course be attained, while even in the short run the flux will cause departures from local equilibrium, selectively depleting some microstates in the reactant zone and enhancing some in the product zone. However, if both reactant and product zones have mean residence times much longer than their internal relaxation times, this selective depletion and enhancement will be negligible, and the approach to global equilibrium will take place without a significant deviation from local equilibrium. The local equilibrium or 'steady-state' approximation is justified whenever the so-called bottleneck really is a bottleneck between the two regions it connects, in the sense of being the chief obstacle to their rapid equilibration. If it is not, then the relaxation process being studied either lacks a clear-cut bottleneck, or else the bottleneck has been incorrectly identified and the true bottleneck lies within the reactant or product zone.

The lack of equilibrium between reactant and product zones leads to a distinctly nonequilibrium distribution in the bottleneck, but fortunately it is one that can be expressed easily (11) in terms of the equilibrium distribution and trajectory information. To do this, the equilibrium probability density $Peq(\underline{p},\underline{q})$ is split into two nonoverlapping parts, $Pa(\underline{p},\underline{q})$ and $Pc(\underline{p},\underline{q})$, the former originating from an equilibrium distribution in $\underline{A}$, the latter from an equilibrium distribution in $\underline{C}$.

For each phase point $(\underline{p},\underline{q})$:
  If the (unique) trajectory through $(\underline{p},\underline{q})$ has been in $\underline{A}$ more recently than it has been in $\underline{C}$, set $Pa(\underline{p},\underline{q})=Peq(\underline{p},\underline{q})$ and set $Pc(\underline{p},\underline{q})=0$.
  Conversely, if the trajectory through $(\underline{p},\underline{q})$ has been in $\underline{C}$ more recently than in $\underline{A}$, set set $Pc(\underline{p},\underline{q})=Peq(\underline{p},\underline{q})$ and set $Pa(\underline{p},\underline{q})=0$.

Since every phase point (except for uninteresting ones accessible from neither $\underline{A}$ nor $\underline{C}$) satisfies one of the two trajectory conditions above and no phase point satisfies both, the two terms add up to the equilibrium density; on the other hand, each term separately represents the situation in which an equilibrium distribution of trajectories attacks the bottleneck from one

side while no trajectories attack from the other side.
The general intermediate case, where A and C are both
populated and internally at equilibrium but out of
equilibrium with each other, can be expressed by saying
that if a nonequilibrium steady state's probability
density is uniformly Xa times the equilibrium value in
A and uniformly Xc times the equilibrium value in C,
then the resulting density in the bottleneck region
will be

$$P_{neq}(\underline{p},\underline{q}) = X_a \cdot P_a(\underline{p},\underline{q}) + X_c \cdot P_c(\underline{p},\underline{q}). \qquad (3)$$

Counting the Trajectories. The generation of trajectories and the estimation of the overall transition
rate are facilitated by defining an arbitrary 6N-1 dimensional dividing surface S in the bottleneck region, and counting the trajectories as they cross
through it.

Figure 2

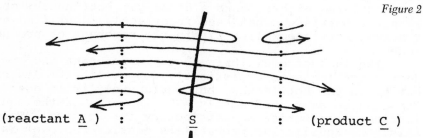

(reactant A)   :   S   :   (product C)

The forward transition rate constant, i.e. the number
of transitions from A to C per unit time and per
unit probability in region Ā, can be expressed generally and rigorously (i.e. assuming only classical mechanics and local equilibrium in A) as

$$W = \frac{\int_{S} d\sigma \ P_{eq}(\underline{p},\underline{q}) \cdot u_{\perp}(\underline{p},\underline{q}) \cdot (u_{\perp}>0) \cdot \xi(\underline{p},\underline{q})}{\int_{A} d\omega \ P_{eq}(\underline{p},\underline{q})}. \qquad (4)$$

Here P_{eq}, the equilibrium probability density defined
earlier, is integrated (dω) over the 6N dimensional
reactant zone A to obtain the normalizing factor in
the denominator. In the numerator, the same density,
is integrated (dσ) over the 6N-1 dimensional surface
S, with various weight factors which, like P_{eq}, are
functions of the coordinates q and momenta p. The
factor $u_{\perp}(\underline{p},\underline{q})$ is the normal component of the veloci-

ty (in 6N space) of the unique trajectory that crosses
the surface S at point (p̲,q̲). It is included because
the crossing frequency through a surface is proportional to the product of local density and velocity; reverse crossings are excluded by the factor (u⊥>0) which
takes the value 1 or 0 according to the sign of
u⊥(p̲,q̲). The integral of the first three factors alone
thus represents the equilibrium forward crossing frequency through the dividing surface, and in early forms
of transition state theory this was usually identified
with the forward transition rate. In fact, because of
multiple crossings, it is only an upper bound on the
transition rate. Multiple crossing trajectories have
been found to be significant in gas phase chemical
reactions (13), and in vacancy diffusion in solids
(14).

   To correct for multiple crossings (and, incidental-
ly for nonequilibrium between reactant and product
zones) Keck (10) and Anderson (11) introduced a third,
trajectory-dependent factor ξ(p̲,q̲) that causes each
successful forward trajectory (i.e. originating in A
and passing through the bottleneck to C) to be counted exactly once, no matter how many times it crosses
S; and causes other trajectories (i.e. those that go
from C to A, from A to A, or from C to C) not to be
counted at all. Many different ξ functions will
achieve this purpose, for example Anderson's:

$$\xi\,(p̲,q̲) = \begin{cases} 1 & \text{if the (unique) trajectory through } (p̲,q̲) \\ & \text{crosses S an odd number of times,} \\ & \text{of which } (p̲,q̲) \text{ is the last,} \\ 0 & \text{otherwise;} \end{cases}$$

or Keck's:

$$\xi\,(p̲,q̲) = \begin{cases} 1/k & \text{if } (p̲,q̲) \text{ is one of the forward crossings} \\ & \text{on a trajectory with k forward} \\ & \text{crossings and k-1 backward crossings,} \\ 0 & \text{otherwise.} \end{cases}$$

In addition to correcting for multiple crossings, the
factor corrects for nonequilibrium between reactant and
product zones, because those parts of S not in equilibrium with A contribute only trajectories for which
the product (u̅⊥>0)·ξ is zero.

   It is clear for topological reasons that the same
value of the transition rate will be obtained regardless of where the dividing surface is placed in B,
provided it intersects all successful trajectories.
Nevertheless, for the sake of better statistics, the

dividing surface should be chosen so as to intersect as
few unsuccessful trajectories as possible. Similarly,
although the two ξ functions have the same mean value,
Keck's appears preferable because it has a smaller variance.

For use, eq. 4 may be rewritten in the form of two
factors, which require somewhat different numerical
techniques for their evaluation:

$$W \;=\; \frac{\displaystyle\int_{\underline{S}} d\sigma\; P_{eq}}{\displaystyle\int_{\underline{A}} d\omega\; P_{eq}} \;\cdot\; < u_{\perp}\cdot(u_{\perp}{>}0)\cdot\xi >_{s} \tag{5}$$

where <>s denotes averaging over an equilibrium ensemble on the surface $\underline{S}$.

The first or 'probability factor' is essentially a
ratio of partition functions, and represents the integrated equilibrium density of phase points on $\underline{S}$ per
phase point in $\underline{A}$. The second or 'trajectory-corrected
frequency factor' is the number of successful forward
trajectories per unit time and per unit equilibrium
density on $\underline{S}$. The ratio of this to the uncorrected
frequency factor $<u_{\perp}\cdot(u_{\perp}{>}0)>_{s}$ represents the number of
successful forward trajectories per forward crossing.
Anderson called this ratio the 'conversion coefficient'
to distinguish it from the 'transmission coefficent' of
traditional rate theory (1), which was usually defined
rather carelessly and given little attention, because
it could not be computed without trajectory information.

Usually one deals with a system whose equations of
motion are invariant under time reversal, and the definitions of the dividing surface and reactant and product regions involve only coordinates, not momenta.
Under these conditions (which will henceforth be as-
sumed) the factor $u_{\perp}\cdot(u_{\perp}{>}0)$ in eqs. 4 and 5 can be
replaced by $\frac{1}{2}|u_{\perp}|$, and the frequency factor (and
conversion coefficient) will be the same in the forward
and backward directions, because every successful forward trajectory is the reverse of an equiprobable successful backward trajectory. One can then use a third
form of the ξ function, viz.

$$\xi(\underline{p},\underline{q}) = \begin{cases} 1/k & \text{if } (\underline{p},\underline{q}) \text{ is any crossing on a trajec-} \\ & \text{tory that makes an odd} \\ & \text{number, } k, \text{ of crossings,} \\ 0 & \text{otherwise.} \end{cases} \tag{6}$$

This function has the least variance of all
ξ functions, because it distributes each trajectory's
weight equally among all its crossings.

When the the activation energy is small compared
to the total kinetic energy, as it is in most systems
with >100 degrees of freedom, the difference between
the microcanonical ensemble and the more convenient
canonical ensemble can usually be neglected. In the
canonical ensemble, the momentum integrals cancel out
of eq. 4, making the probability factor a simple ratio
of configurational integrals. Combining this with the
time-reversal-invariant form of the frequency factor
and the optimum ξ function of eq. 6, we get

$$W \; = \; \frac{Q^{\ddagger}}{Qa} \cdot < \tfrac{1}{2} \, |u_{\perp}(\underline{p},\underline{q})| \cdot \xi(\underline{p},\underline{q}) \; >s, \tag{7}$$

where Qa and $Q^{\ddagger}$ are integrals of $\exp(-U(\underline{q})/kT)$ over,
respectively, the 3N dimensional reactant region and
the 3N-1 dimensional dividing surface in configuration
space. This exact expression for the transition rate
is the one that will be used most often in the remain-
der of this paper.

Definition of a Successful Transition. It is
clear that the transition rate depends on the boundar-
ies adopted for the bottleneck region $\underline{B}$, which a trajectory must traverse to be counted as successful. If
$\underline{B}$ is made very narrow, the transition rate will be ov-
erestimated, because dynamically-correlated multiple
crossings will be counted as independent transitions;
on the other hand, if $\underline{B}$ is enlarged to include all of
configuration space, trajectories will never leave B
and the transition rate will be zero. However, if the
assumed bottleneck indeed represents the chief obstacle
to rapid equilibration between two parts of configuration space, there will be a range sizes over which the
transition rate is nearly independent of the definition
of $\underline{B}$. These 'reasonable' definitions will make B small
enough to exclude most of the equilibrium probability,
yet large enough so that a trajectory passing through $\underline{B}$
in either direction is unlikely to return through $\underline{B}$
immediately in the opposite direction.

The time of return can itself be made the criter-
ion of success, by forgetting about the $\underline{B}$ region and
counting two consecutive crossings of $\underline{S}$ as independent
transitions if and only if they are separated by a time
interval greater than some characteristic time To,
e.g. the autocorrelation time of the velocity normal to

the dividing surface. A successful transition, then,
is a portion of trajectory that crosses $\underline{S}$ an odd number
of times, at intervals less than To, preceded and
followed by crossing-free intervals of at least To.
This criterion of success emphasizes the fact that unless the mean time between transitions is long compared
to other relaxation times of the system, successive
transitions will be correlated, and the transition rate
will be somewhat ill-defined. Such correlated transitions, representing a breakdown of the random walk hypothesis, are significant in solid state diffusion
(14,15), at high defect jump rates. The correlations
may be investigated either by simulating the system
directly, without bottleneck methods, or by continuing
trajectories started in the bottleneck far enough forward and backward in time to include any other transitions correlated with the original one.

Sampling the Equilibrium Distribution in the Bottleneck. In order to generate representative trajectories and evaluate the corrected frequency factor, one
needs a sample of the equilibrium distribution
$Peq(\underline{p},\underline{q})$ on the surface $\underline{S}$, where the total equilibrium
probability is very low. For very simple systems (13)
this sample can be generated analytically, but for anharmonic polyatomic systems it can only be obtained
numerically, by doing a molecular dynamics or Monte
Carlo machine experiment designed to sample the equilibrium distribution on $\underline{S}$ correctly, while greatly enhancing the system's probability of being on or near
$\underline{S}$. This may be accomplished by a Hamiltonian of the
form

$$H^*(\underline{p},\underline{q}) = \begin{cases} H(\underline{p},\underline{q}) & \text{if } (\underline{q}) \text{ is within a small} \\ & \text{distance } \delta \text{ of } \underline{S}, \\ +\infty & \text{otherwise.} \end{cases} \qquad (8)$$

One can do dynamics under this Hamiltonian by making
the trajectory undergo an elastic reflection whenever
it strikes one of the infinite barriers (14). Under
H*, the different parts of S would be visited with the
same relative frequency as in an unconstrained equilibrium machine experiment, but with a much greater absolute frequency; thereby allowing a representative sample of, say, 100 representative points on S to be as-
sembled in a reasonable amount of computer time. If
the equilibrium distribution is canonical the momentum
distribution will be Maxwellian and independent of
coordinates; hence, representative points $(\underline{p},\underline{q})$ can
be generated by taking $\underline{q}$ from an equilibrium Monte

Carlo run constrained to make moves on the   3N-1 dimensional dividing surface in configuration space, and
supplying momenta from the appropriate multidimensional
Maxwell distribution. Alternatively, the dividing surface may be sampled by an unconstrained Monte Carlo run
that is encouraged to remain near $\underline{S}$ by adding to the
potential a holding term that is constant on $\underline{S}$ but
increases rapidly as $\underline{q}$ moves away from $\underline{S}$.

   A well-chosen dividing surface $\underline{S}$ should satisfy
these three criteria: 1) its conversion coefficient
should not be too small, 2) its definition should be
simple enough to be implemented as a constraint or
holding term in a MC or MD run, and 3) the autocorrelation time of this run should not be too large. If the
bottleneck technique is to represent any saving over
straight simulation, the total machine time expended
per statistically-independent successful transition
(including time to generate a statistically-independent
starting point on $\underline{S}$, time to compute the trajectory
through it, and overhead from unsuccessful trajectories) must be less than the mean time between spontaneous transitions in a straightforward non-bottleneck
simulation. Ordinarily, if the bottleneck is a single
compact region in configuration space, it will not be
difficult to find a dividing surface that satisfies all
three criteria. On the other hand, if the bottleneck
is broad and diffuse, containing many parallel independent channels, the only surfaces that satisfy the
first criterion may be so complicated and hard to define that they fail the second and third (In this connection it should be noted that the 'continental
divide' or 'watershed' between two reservoirs, which
might appear an ideal dividing surface because of its
high conversion coefficient, is not usable in practice
because it is defined by a nonlocal property of the
potential energy surface). Fig. 3 suggests a broad,
diffuse bottleneck whose watershed (dotted line) is so
broad and so contorted that no simple approximation to
it can have a good conversion coefficient.

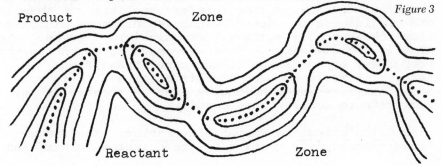

Product   Zone   *Figure 3*

Reactant   Zone

It is not known whether such pathological bottlenecks
occur in practice.

One important kind of broad bottleneck, probably
not pathological, is found in chemical reactions in
liquid solutions; where most of the solvent molecules
are geometrically remote from, and therefore only weak-
ly coupled to, the atoms immediately involved in the
transition. The remote atoms exert only a mild perturbing effect on the transition, and need not be in
any one configuration for the transition to occur. In
other words, if a number of trajectories for successful
transitions were compared, all would pass through a
single small bottleneck in the subspace of important
nearby atoms, but the same trajectories, when projected
onto the subspace of remote atoms, would not be concen-
trated in any one region. In the full configuration
space, the bottleneck will therefore appear broad and
diffuse in the directions of the weakly-coupled degrees
of freedom.

The obvious approach to this problem is to look
for a dividing surface in the subspace of strongly
coupled 'participant' degrees of freedom, for which the
bottleneck is well localized. In the directions of the
weakly-coupled 'bystander' degrees of freedom, the watershed is broad and diffuse; but one can reasonably
hope that--precisely because of this weak coupling--it
is not highly contorted in these directions, and that
therefore the surface $\underline{S}$ will be a good approximation
to it. Of course it may not always be easy identify
the participants and bystanders correctly.

The problem of separating the participants from
the bystanders has come up in attempts to simulate dissociation of a pair of oppositely charged ions in water
(16). If the dividing surface is taken to be a surface
of constant distance between the two ions, the trajectory typically recrosses this surface many times without making noticeable progress toward dissociation or
association. This appears to be because of a constraining cage of water molecules around the ions,
which must rearrange itself before the ions can associ-
ate or dissociate. Nevertheless, spontaneous dissocia-
tions occasionally occur rather quickly. This suggests
that if the dividing surface were made to depend in the
proper way on the shape of the cage, transitions
through it would be much less indecisive. It is not
known how many water molecules must be treated as participants to achieve this result.

Calculating the Probability Factor. The transitions generated by continuing trajectories forward and

backward in time from starting points on S will be representative of spontaneous transitions through the
bottleneck, but the absolute transition rate will not
yet be known, because the first factor of eqs. 5 and 7
is not known, and cannot be computed from information
collected in the bottleneck region alone. This factor
is the Boltzmann exponential of the free energy difference (or for a microcanonical ensemble, entropy difference) between a system constrained to the reactant region and a system constrained to the neighborhood of
the dividing surface. For very simple or harmonic systems the free energy difference can be calculated analytically, but in general, it can only be found by
special Monte Carlo or molecular dynamics methods.
These methods resemble the calorimetric methods by
which free energy differences are determined in the
laboratory, in that they depend on measuring the work
necessary to conduct the system along a reversible path
between the two macrostates, or between each of them
and some reference macrostate of known free energy.
Laboratory calorimetry measures free energy as a function of independent state variables like temperature.
Machine experiments are less limited: they can measure
the free energy change attending the introduction of an
arbitrary constraint or perturbing term in the Hamiltonian. In the present case, for example, one could measure the reversible work required to squeeze the system
from the reactant zone into the neighborhood of S by
integrating the pressure of collisions against one of
the constraining barriers of H*, as it is moved slowly
into place.

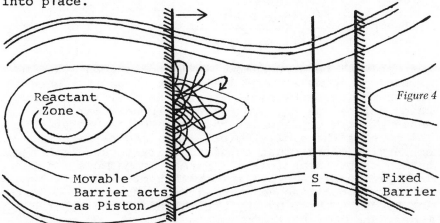

Reactant
Zone

Movable
Barrier acts
as Piston

S

Figure 4

Fixed
Barrier

Alternatively, one could measure the reversible work
along a path between the bottleneck and one reference
system (e.g. a quadratic saddle point), and along

another path between the reactant zone and a second
reference system (e.g. a quadratic minimum), and sub-
tract these. Computer calorimetry is easiest to perform in the canonical ensemble, where any derivative of
the free energy is equal to the canonical average of
the same derivative of the Hamiltonian, measurable in
principle by a Monte Carlo run:

$$\partial(A/kT)/\partial\lambda \;=\; <\, \partial(H/kT)/\partial\lambda \,> \tag{9}$$

Here A is the Helmholtz free energy, λ is an arbitrary parameter of the Hamiltonian, and $<>$ denotes a
canonical average. For more information about
'computer calorimetry' see refs. 17, 18, and 19.

    Relation of Exact TST to the Harmonic Approxima-
tion. In the canonical ensemble, the most familiar TST
expression for the rate constant is probably

$$W \;=\; \frac{kT}{h} \cdot \frac{Z\ddagger}{Za} \cdot K \;, \tag{10}$$

where Za and $Z\ddagger$ are dimensionless quantum or classical partition functions of the A-constrained and
S-constrained systems, calculated with respect to the
same energy origin, and K is a transmission coefficient. This equation is exact and equivalent to eq. 7
if the partition functions are computed classically,
and if K is taken to be the conversion coefficient,

$$K = \bar{\xi} = \frac{<|u_\perp(\underline{p},\underline{q})| \cdot \xi(\underline{p},\underline{q})>s}{<|u_\perp(\underline{p},\underline{q})|>s}, \tag{11}$$

but, as will be seen below, it is not a good quantum
mechanical formula. Eq. 10 is most frequently used in
the harmonic approximation, with the dividing surface $\underline{S}$
being defined as the hyperplane perpendicular (in
mass-weighted configuration space) to the unstable normal mode at the saddle point. This choice makes the
conversion coefficient equal to unity because (in the
harmonic approximation) all normal modes move independently; therefore a trajectory that crosses this hyperplane with positive velocity in the unstable mode cannot be driven back by any excitation of the other
modes.
    The partition functions $Z\ddagger$ and Za are also easi-
ly evaluated in the harmonic approximation from products of the stable normal mode frequencies at the sad-

dle point and minimum, respectively. One thus obtains
a formula expressing the transition rate in terms of
local properties at two special points of the potential
energy surface-- the minimum of the reactant zone, and
the saddle point in the bottleneck:

$$
W = \frac{\displaystyle\prod_{}^{3N} \nu_{min}}{\displaystyle\prod_{}^{3N-1} \nu_{sp}} \cdot \exp(\ -(Usp-Umin)\ /\ kT\ ), \tag{12}
$$

where ν_{min}, and ν_{sp} denote the stable normal mode
frequencies, and $Umin$ and Usp denote the potential
energy, at the minimum and saddle point, respectively.
The system is assumed to have no translational or rotational degrees of freedom.

This traditional, and still very useful, form of
transition state theory is valid whenever quantum effects are negligible and the potential energy surface
is quadratic for a vertical distance of several kT
above and below the saddle point and minimum. Aside
from assuring the accuracy of the harmonic partition
functions, the latter condition justifies setting
$\bar{\xi}$ = 1 by assuring that trajectories crossing the saddle point hyperplane will not be reflected back until
they have fallen several kT below the saddle point
energy. In practice, although it is hard to prove
(20), this makes multiple crossings very unlikely
(21).

Much of the power of eq. 12 comes from the existence of powerful, locally-convergent methods for finding energy minima and saddle points, and methods for
evaluating products of normal mode frequencies. Depending on the number of degrees of freedom, variable
metric (22) minimizers like Harwell Subroutine VA13A
or conjugate-gradient (23) minimizers like VA14A converge to the local energy minimum much faster than the
obvious method of damped molecular dynamics. Saddle
points can be found (24) similarly by minimizing the
squared gradient $|\nabla U|^2$ of the energy (the starting
point for this minimization must be fairly close to the
saddle point, otherwise it will converge to some other
local minimum of $|\nabla U|^2$, such as an energy minimum
or maximum). Once the saddle point has been found,
existing routines, taking advantage of the sparseness
of $\nabla\nabla U$ for large n, are sufficient to extract the
unstable mode at the saddle point and compute the pro-

duct of stable mode frequencies (essentially the determinant of $\nabla\nabla U$) even for systems with several hundred
atoms.

Even when the harmonic approximation is not quan-
titatively justified it provides a convenient starting
point for exact treatments. Thus, even if the potential energy surface is anharmonic in the bottleneck, it
is often smooth enough for there to be a principal sad-
dle point that can be found by minimizing $|\nabla U|^2$.
The harmonic hyperplane through this saddle point often
makes a good dividing suface, through which most crossings lead to succeed. Similarly, the harmonic configurational integral on the hyperplane is a good starting
point for a calorimetric Monte Carlo determination of
the exact configurational integral on the same hyperplane. It may be necessary to restrict the hyperplane
laterally, to avoid irrelevant portions of it that may
extend beyond the bottleneck region.

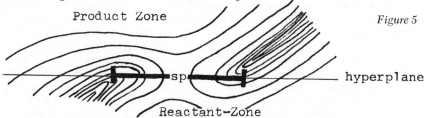

Product Zone   *Figure 5*

sp   hyperplane

Reactant-Zone

The single-occupancy constraints mentioned on page 90
of ref. 14 are an example of such lateral restric-
tion).

In systems whose bottlenecks are diffuse because
of weakly-coupled 'bystander' degrees of freedom, it
may be useful to look for a saddle point and harmonic
hyperplane in the subspace of strongly coupled
'participant' degrees of freedom, e.g. by minimizing
$|\nabla U|^2$ with respect to the participants while the bystanders are held fixed in some typical equilibrium
positions. In general, minimum and saddle-point seek-
ing routines will be useful whenever the potential energy surface (or its intersection with the subspace of
participants) is smooth--i.e. free of numerous small
wrinkles and bumps of height kT or less. When such
roughness is absent, the typical bottleneck will not
contain many saddle points.

Quantum Corrections. The obvious way to introduce
quantum corrections in eq. 10 would be to interpret Z_a
and $Z\ddagger$ as quantum partition functions; however, this
neglects tunneling ($Z\ddagger$, being the partition function of
a system constrained to the top of the activation bar-

rier, knows nothing about the barrier's thickness). In the harmonic approximation tunneling can be included as a 1-dimensional parabolic barrier correction, which has the same magnitude (but opposite sign) as the lowest-order quantum correction to the partition function of a parabolic well of the same curvature (25, 26). This means that, in the harmonic approximation and to lowest order in h, the classical transition rate is multiplied by a factor depending only on the sums of squares of the normal mode frequencies at the saddle point and minimum:

$$\frac{W \text{ quantum}}{W \text{ class.}} = 1 + \frac{1}{24}\left(\frac{h}{kT}\right)^2\left[(\sum^{3N}\nu^2_{min}) - (\sum^{3N}\nu^2_{sp})\right]. \quad (13)$$

The unstable mode at the saddle point has an imaginary frequency, and contributes negatively to the second sum, raising the transition rate. When this correction is applied to eq. 13, one still has an expression for the rate in terms of purely local properties at the saddle point and minimum. The size of this rather readily-calculated lowest-order correction can serve as a guide to whether more sophisticated quantum corrections are necessary.

The conditions for validity of the harmonic approximation in eq. 13 (i.e. that the potential be quadratic within a few de Broglie wavelengths $h/\sqrt{2\pi mkT}$ in all directions from the saddle point) are somewhat opposed to its conditions of validity in eq. 12 (i.e. that the potential be quadratic within a few kT above and below the saddle point), and for some chemical reactions, particularly those involving hydrogen, the harmonic approximation is not justified quantum mechanically in the temperature range of interest (27) even though it would be classically (21). For these reactions, more sophisticated 1-dimensional tunneling corrections to eq. 10 usually also fail, and it becomes necessary to use a method that does not assume separability of the potential in the saddle point neighborhood.

Such a method has recently been developed by Miller, et. al. (28). It uses short lengths of classical trajectory, calculated on an upside-down potential energy surface, to obtain a nonlocal correction to the classical (canonical) equilibrium probability density $P_{eq}(\underline{p},\underline{q})$ at each point; then uses this corrected density to evaluate the rate constant via eq. 4. The method appears to handle the anharmonic tunneling in the reactions H+HH and D+HH fairly well (28), and can

be applied economically to systems with arbitrarily
many degrees of freedom.

Another quantum problem, the wide spacing of vibrational energy levels compared to kT, has caused trouble in applying bottlneck methods to simple gas phase
reactions (29), making them sometimes less accurate
than 'quasiclassical' trajectory calculations in which
trajcetories are begun in the reactant zone with quantized vibrational energies. This problem should be
much less severe in polyatomic systems, because of the
closer spacing of energy levels.

Systems with Many Bottlenecks.

So far we have considered a system with two reservoirs
separated by one bottleneck; in general a polyatomic
system will have many reservoirs in its configuration
space, and the location of the critical bottleneck or
bottlenecks will be unknown. Here we will first distinguish critical and rate-limiting bottlenecks from
less important ones, and then discuss several more or
less heuristic methods for for finding bottlenecks.

Definition of Critical and Rate-Limiting Bottlenecks. The hypothesis of local equilibrium within the
reservoirs means that the set of transitions from reservoir to reservoir can be described as a Markov process without memory, with the transition probabilities
given by eq. 4. Assuming the canonical ensemble and
microscopic reversibility, the rate constant W_{ji}, for
transitions from reservoir i to reservoir j can be
written

$$W_{ji} = \exp -\left(\frac{B_{ji} - A_i}{kT}\right) \tag{14}$$

where

$$A_i = -kT \ln Q_i \tag{15}$$

is the free energy of reservoir i, and

$$B_{ij} = B_{ji} = -kT \ln (Q\ddagger \cdot < \frac{1}{2} |u_\perp(\underline{p},\underline{q})| \cdot \xi(\underline{p},\underline{q}) > s) \tag{16}$$

is a symmetric 'free energy' of the bottleneck, with
$Q\ddagger$ and $< >s$ being the configurational integral and
equilibrium expectation on the dividing surface between
reservoirs i and j (equations 15 and 16 fix the origin of the free energy scale by defining the A's and
B's microscopically in terms of configurational integ-

rals; however a consistent set of A's and B's could be
defined macroscopically from the Wij, by arbitrarily
setting one of the A's to zero and solving eq. 14 recursively for the B's and the other A's). The system
of reservoirs and bottlenecks can be represented on an
'activation energy diagram' with valley-heights given
by the A's and peak-heights given by the B's.

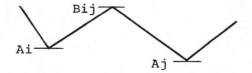

Figure 6

Microscopic reversibility of the equations of motion is important-- without it the Bij would not be
symmetric, and the relative occupation probabilities of
reservoirs i and j in the long-time limit, here given
by $P_i/P_j = W_{ij}/W_{ji} = \exp((A_j-A_i)/kT)$, could no
longer be expressed in terms of local properties of the
two reservoirs alone, but would depend on all paths
connecting them.
The abscissa in activation energy diagrams is always somewhat arbitrary; the ordinate, although it cannot be assigned a definite meaning in the general chem-
ical-kinetic situation of coupled reactions of differ-
ing order (30), has the exact meaning given in eq. 14
when the transitions are defined, as they are here, by
a microscopically reversible set of first order rate
constants. If there are many interconnecting reservoirs, the peak and valley representation becomes in-
convenient, and the system is better represented as an
undirected graph whose vertices are the reservoirs and
whose edges are the bottlenecks.
Since free energies tend to be large compared to
kT, it is reasonable to assume that no two reservoirs
have the same free energy to within kT, and that no two
bottlenecks do either. Under these conditions, exit
from any reservoir is overwhelmingly likely to occur
through the lowest bottleneck leading out, and given
any two reservoirs x and y , there is a well-defined
set of reservoirs and bottlenecks which the system will
probably visit on its way from x to y. This set consists simply of all the places that would get wet if
water were poured into x until it began running into
y (cf. fig. 7).

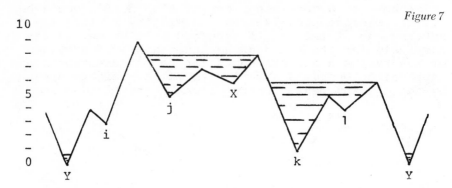

Figure 7

In fig. 7, the wet set consists of all the reservoirs
except i , and all the bottlenecks except ji and iy.
The reservoir y is shown twice to avoid having to
superimpose visually the two parallel paths (x-k-l-y)
and (x-j-i-y) that lead from x to y.
      The hydrological construction leads to a descend-
ing sequence of lakes, each comprising a set of elemen-
tary reservoirs that reach a common local equilibrium
and look from the outside like a single reservoir. The
mean residence time for a lake is the positive exponential of its depth (the depth of a compound lake is simply the depth of its deepest part, compared to which
all other parts are negligible, because of the rule
that the A's typically differ by more than kT). Of
the bottlenecks that are visited, the submerged ones
like xj and kl are typically visited many times,
and have hardly any influence on the mean time required
to get from x to y.   The critical bottlenecks are
ones like xk and ly that stand at the spillways of
lakes. The system typically passes through each critical bottleneck exactly once. One of the critical bottlenecks, the one with the deepest lake behind it, is
rate-limiting: most of the time is spent waiting in
that lake.
      (It is sometimes wrongly supposed that the highest
bottleneck, here xk, is rate-limiting; in fact
bottleneck ly is, because its lake is deeper than
that behind xk. The highest bottleneck is thus
'path-determining' without necessarily being rate-limiting. The complicated relation among rates and bottlenecks is shown by the fact that if bottleneck xk
were raised, so that the jx lake overflowed to the
left instead of to the right, the mean time to pass
from x to y would actually be decreased, because the
deepest lake would have a depth of four instead of
five.)

Finding the Bottlenecks. In order to carry out
the hydrological construction, one must be able, given
a reservoir i and a list of the n lowest bottlenecks leading out of it, to find the next lowest, its
transition rate Wji and the new reservoir j that it
leads to. A straightforward MD or MC simulation would
eventually find all the relevant bottlenecks and reservoirs, but only at the cost of waiting thousands of
years in the deep lakes, which is precisely what we are
trying to avoid. There are several ways of getting a
polyatomic system to escape from one local minimum or
reservoir into another; but unfortunately none of them
can be trusted to escape via the bottleneck of lowest
free energy, as is required for the hydrological construction. Therefore they must be used rather conser-
vatively, in an attempt to gradually discover and fill
out the unknown graph of reservoirs and transition
rates, without missing any important bottleneck.
    Escape methods are most powerful when used in con-
nection with static energy minimization and
saddle-point finding routines, in an effort to cata-
logue all the relevant saddle points and minima on the
potential energy surface. This approach should be used
whenever the potential energy surface is smooth on a
scale of kT, so that the typical barrier height between adjacent local minima is high enough to justify
treating each local minimum as a separate reservoir and
each saddle point as a separate bottleneck. The main
static methods of escape are 1) systematic search, 2)
intuition, 3) normal mode thermalization, and 4)
'pushing'.
    1) Systematic search of the neighborhood. This is
practical only if the search is conducted in a subspace
of low dimensionality, because the number of mesh
points required grows exponentially with the dimensionality. It is usually advisable, at each mesh point, to
locally minimize the energy with respect to all the
degrees of freedom not being searched. This is called
'adiabatic mapping'. A systematic search with sufficiently fine mesh in all relevant degrees of freedom
will indeed locate all saddle points and minima in a
given neighborhood, but it is usually prohibitively
expensive. Methods that do not search everywhere are
in principle unreliable because it is possible for a
saddle point or minimum on the potential energy surface
to be so sharply localized that it is undetectable a
short distance away. (This may seem to contradict the
notion that since every saddle point has a unique 1-di-
mensional gully or steepest-descent path connecting it

to each of two minima, it ought to be possible to follow the gullies from minimum to saddle point to minimum
all over the potential energy surface. Unfortunately,
neither these gullies nor the 3N-1 dimensional wat-
ersheds between adjacent minima are locally-definable
properties of the potential energy surface).

2) Intution: considerations of symmetry and common
sense (aided perhaps by model building) often make the
approximate locations of the relevant minima and saddle
points obvious.

3) Levitt and Warshel (31,32) have used an method
called 'normal mode thermalization' to simulate stati-
cally the effect of heating to a temperature above the
barrier height between adjacent minima. Starting at a
one local minimum, the system is displaced along each
normal mode by an amount that would correspond to kT
energy rise on the local quadratic approximation to the
potential energy surface; however, the 'temperature'
used is so high that on the real potential energy surface the system is displaced out of its original watershed, and subsequent energy minimization leads to a
new local minimum, from which the whole process can be
repeated. Like explicit heating, this method preferentially displaces the system in the easy directions--
i.e. along the softer normal modes-- which are less
likely to produce immediate atom-atom overlaps.

4) 'Pushing'. This consists minimizing the energy
of a system in which the original minimum has been de-
stabilized by an artificial perturbing term in the potential energy. Such pushing potentials have been used
in energy minimization studies on proteins by Gibson
and Scheraga (33) and by Levitt (32), and are quite
similar in spirit to the methods used by Torrie and
Valleau (19) to push Monte Carlo systems into desired
regions of configuration space.

In the case of energy minimization, the goal of
the added term should be to make what was a local minimum flat, or slightly convex, thus causing the system
to roll away to another minimum. The obvious term to
do this is a paraboloidal mound complementary in shape
to the harmonic neighborhood of the local minimum:

$$U'(\underline{q}) \quad = \quad U(\underline{q}) \quad + \text{Upush}(\underline{q}), \qquad \text{where}$$

$$\text{Upush}(\underline{q}) \quad = \quad -(\underline{q}-\underline{q}min)\cdot\nabla\nabla U(\underline{q}min)\cdot(\underline{q}-\underline{q}min), \qquad (17)$$

with $\underline{q}min$ denoting the coordinates of the minimum.
One may also define a spherically symmetric pushing
potential,

$$\text{Upush}(\underline{q}) = -\text{const} \; |(\underline{q}-\underline{q}\text{min})|^2. \tag{18}$$

The potential of eq. 17 pushes the system away from the original minimum in the directions of negative deviation from harmonicity. The spherically symmetric potential of eq. 18 pushes the system away preferentially along the directions of low curvature. The pushing potentials used by Levitt (32) were of the symmetric type and incorporated a smooth cutoff at a range of several atomic diameters; this is avoids having the pushing potential dominate the energy at large distance, seriously distorting any new minimum the system escapes into. More recently (34) Levitt has used unsymmetrical pushing potentials.

Since pushing potentials are not guaranteed always to escape via the lowest saddle point, it would be wise to use them systematically in an effort to find all the easy escapes from the given initial minimum. This can be done by repeating the escape minimization several times, each time adding to the potential a short-ranged repulsive term placed so as to obstruct the pervious escape route.

Having escaped from one local minimum to an adjacent one, the next task is to find the saddle point, choose a good dividing surface and calculate the transition probabilities W_{ij} and W_{ji}. If escape was achieved by pushing, the escape path typically passes through the bottleneck region, and the highest point (i.e. the point having highest <u>unperturbed</u> energy) on this path is often close enough to the saddle point to serve as a starting point for a locally convergent minimization of $|\nabla u|^2$, to find the saddle point. Once the saddle point has been found, the unstable mode and perpendicular hyperplane may be constructed in the usual manner.

If no escape path is available (e.g. if the second minimum were known a priori by reasons of symmetry or if it were found by a systematic search), an escape path can be generated by the 'push-pull' method. This is like pushing, except that it supplements the pushing potential in the minimum one wishes to leave with an attractive 'pulling' potential in the minimum one wishes to enter. The strengths and ranges of these potentials are gradually increased until the desired transition occurs. Tests of the push-pull method (35) on chirality reversal of a 10-atom model polymer showed it superior to the common method of one-dimensional constrained minimization, which did not come close enough to the saddle point to begin a convergent minimization of the squared gradient. The pitfalls of sad-

dle-point finding methods based on constrained minimization have been noted by McIver and Komornicki (24)
and Dewar and Kirschner (36).

When the potential energy surface is rough on the
scale of kT, so that local minima are very numerous and
separated by barriers of height kT or less, energy
minimization methods are not very helpful, and it becomes necessary to use escape methods that will enable
a finite-temperature MC or MD system to escape from a
reservoir containing many local minima, through a bottleneck perhaps containing many saddle points. Aside
from intuition, there are two basic methods: 1) heating, and 2) pushing.

1) Heating--a MC or MD system can always be induced
to leave a restricted region in configuration space by
raising its temperature or equivalently by arbitrarily
making the atoms smaller or softer. Heating has the
disadvantage of favoring escape via a wide bottleneck
regardless of its height on the potential energy surface; this may not be the bottleneck having lowest free
energy at the temperature of interest.

2) Pushing can be best be applied to a MC and MD
system if one has in mind a reaction coordinate,
$r(\underline{q})$, i.e. some function of the coordinates $\underline{q}$ that,
because it takes on a rather limited range of values,
suggests that the system is trapped in a rather limited
part of configuration space. A Monte Carlo run under
the unperturbed potential U would yield a fairly narrow distribution of values of r, representable as a
histogram, h(r):

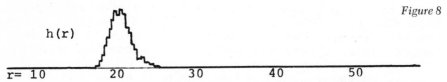

Figure 8

Suppose one is interested (as Torrie and Valleau were)
in the equilibrium probability of an r value, say
r=30, outside the observed range; alternatively, one
may suspect that p(r), the true equilibrium distribution of r, is bimodal, with another peak around
r=40, but that a bottleneck around r=30 is prevent-
ing this peak from being populated.

In order to push the local equilibrium ensemble
out of the range r=15-25, it suffices to perform a
Monte Carlo run under the potential

U' = U + Upush, with

Upush($\underline{q}$) = +kT ln f($r(\underline{q})$),   (19)

where f(r) is an always-positive function chosen to
approximate the histogram h(r) in the range 15-25
where data have been collected and to be a reasonable
extrapolation of h(r) in the region where data are
desired but none have been collected.
    It is easy to show that the equilibrium distribu-
tion under the perturbed potential U' is related to
that under U by

$$\frac{p'(r)}{p(r)} = \frac{Q}{Q'} \frac{1}{f(r)},$$ (20)

where Q and Q' denote the two systems' configurational
integrals. The histogram h' obtained under U' will
thus be approximately flat where h was peaked, and
will extend at least slightly into the range not visited by h.

Figure 9

h'(r)

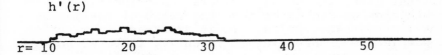

r= 10   20   30   40   50

If there is a bottleneck at r=30, the system is much
more likely to find it and suddenly leak through; if
not, one has a least measured the equilibrium distribution of r in a region where it would be too low to
measure directly. The normalizing factor Q/Q', necessary to make the connection between p and p', can
found be from the histograms via eq. 20 or, more accurately, by eqs. 12a and 12b of reference 17.
    If the system suddenly and irreversibly leaks into
the region around r=40, indicating a bottleneck, the
function f(r) should be revised to flatten out both
peaks of the bimodal distribution, and produce an approximately uniform distribution over the whole range
r=20 to 40. Sampling this flattened-out ensemble
serves two purposes:
    1) It allows a representative sample of configura-
tions on the dividing surface to be collected in a reasonable amount of computer time (the dividing surface
is conveniently defined by r(q) = rmin, where rmin
is the minimum of the bimodal distribution of p(r)

that would obtain under the unperturbed potential   U.)
From these, trajectories can be calculated in the usual
way, to obtain the second factor of eq. 7.
   2) By virtue of the known relation (eq. 20) between p
and p', it establishes a calorimetric path connecting
the reactant region with the bottleneck, allowing the
first factor in eq. 7 to be calculated.
   The definition of the reactant coordinate used in
the MC pushing method may be derived from a separation
into 'participant' and 'bystander' degrees of freedom,
or it may be arrived at intuitively or empirically.
Generally speaking, the more cleanly a reaction coordinate separates the two peaks of a bimodal distribution,
the higher the conversion coefficient that can be achieved with it.

   Speeding up the Sampling of Configuration Space.
Bottleneck methods allow infrequent events to be simu-
lated with very little explicit dynamical calculation,
since the trajectory only needs to be followed forward
and backward until it leaves the bottleneck. On the
other hand, particularly for strongly anharmonic systems, they demand a great deal of MC or MD sampling of
constrained or biased equilibrium ensembles, viz. the
ensemble on the dividing surface, the ensemble in the
reactant zone, and perhaps several calorimetric in-
termediates needed to compute the ratio of configurational integrals, $Q\ddagger/Qa$. It is important to be able
to sample these ensembles efficiently, i.e. without
expending too much computer time per statistically-independent sample point. This section discusses several
curable kinds of slowness commonly encountered in equilibrium sampling. The simplest kind of slowness, and
perhaps the most serious, is due to an unrecognized
bottleneck within the one of the equilibrium ensembles. If the unrecognized bottleneck is fairly easy to
pass through, it will only increase the autocorrelation
time of the run sampling the ensemble; if it is hard,
it will lead to a completely erroneous sample. The
cure is to find the bottleneck and treat it explicitly.
   Another kind of slowness comes from the approximately 1000-fold disparity between bonded and nonbonded
forces among atoms. This means that a typical covalent
bond undergoes about 30 small-amplitude, nearly-harmon-
ic vibrations in the time required for any other significant molecular motion to take place. In doing dynamics calculations, these fast vibrational modes are a
nuisance because they force the use of a very short
time step, about .001 psec. or less. Fortunately, they

can be gotten rid of in either of two ways: 1) they can be artificially slowed down (without affecting the equilibrium statistical properties of the system) by, in effect, giving them extra mass (37); 2) they can be frozen out entirely by incorporating constraints on bond distances and angles in the equations of motion. It was only recently recognized (38) that such constraints, even when applied to a large number of bonds simultaneously, need not appreciably increase the machine time required to do one integration step. Of course the mass-modified system does not have the same dynamics as the original system, and the rigid-bond system has neither the same dynamics nor the same statistical properties; however, accurate dynamics is needed only in the bottlenecks-- correct statistical properties are sufficient elsewhere. In view of the near-harmonicity of the bonded vibrations, it is probable that their effect on the statistical properties could be computed as a perturbation to the statistical properties of a rigid-bond system.

A third kind of slowness, that due to hydrodynamic modes, has been discussed already. It is difficult to do anything about these slow collective modes, but fortunately they cannot cost very many orders of magnitude in a system of a few thousand atoms or less.

A final kind of slowness is that which sometimes arises (39, 17) in Monte Carlo sampling under a biased potential of the form of eq. 19. Sometimes these runs exhibit discouragingly long autocorrelation times for diffusion of the reaction coordinate back and forth along its artifically broadened spectrum. The reason for this is not always clear, but sometimes it may be due to a strong gradient of energy and entropy parallel to the reaction coordinate, so that one end of the spectrum represents a small, low-energy region of configuration space while the other end represents a large region of uniform, moderately-high energy. Ordinary Monte Carlo transition algorithms (8), which make trial moves symmetrically in configuration space and then accept or reject them according to an energy criterion, cannot move very efficiently in such a gradient, because most trial moves are made in the direction of increasing entropy, only then to be rejected for raising the energy. This problem might be ameliorated by using an unsymmetrical Monte Carlo transition algorithm, one that made trial moves more often in directions suspected of leading toward the small, low-energy region, and compensated for this bias by giving a one-way energy reward to moves in the opposite direction.

Summary.

Some phenomena occurring in systems of 3 to 10,000 atoms are so infrequent that they would take thousands of years to simulate on a computer. Such long time phenomena (many orders of magnitude longer than the microscopic system's longest hydrodynamic relaxation time) involve a bottleneck or activation barrier, which, if it can be discovered, can be used to speed up the simulation by many orders of magnitude. The machinery for doing this consists of transition state theory supplemented by classical trajectory calculations to correct for multiple crossings and by 'calorimetric' Monte Carlo methods to evaluate analytically intractable partition functions.

Before the development of the digital computer, the main weakness of transition state theory was its dependence on the harmonic approximation; now its main weakness, and its main potential for future improvement, is in algorithms for finding bottlenecks.

When the energy surface is smooth on a scale of kT, bottlenecks can be identified with saddle points, and the need is for an algorithm that, given a potential minimum, will find all the reasonably low saddle points leading out of it. Existing algorithms are unreliable in principle (because a saddle point may be invisible a short distance away), but may be reliable in practice. More empirical testing of them is needed.

When the potential energy is rough on a scale of kT, saddle points (and their convenient unstable-mode hyperplanes) are no longer a good guide, and the job of selecting the reaction coordinate and dividing surface becomes much more arbitrary and empirical. An important and poorly-understood intermediate case is a potential energy surface that is smooth in some directions (the 'participant' degrees of freedom) and rough in other directions (the 'bystander' degrees of freedom).

Table I. outlines the steps for finding the bottleneck, evaluating the rate constant, and generating typical trajectories for infrequent events.

Table I. Flowchart for Bottleneck Simulation of Infrequent Events.

Flowchart Step	Harmonic (a)	Smooth (b)	Smooth only for participants (c)	Rough for all
(Potential Energy Surface $U(q)$)				
Characterize Reactant Zone A	Local minimum of the potential energy U	Local minimum of the potential energy U	Equilibrium MC or MD run	Equilibrium MC or MD run
Escape from A and find bottleneck or saddle point	Static Push to escape. Minimize $\lvert\nabla U\rvert^2$ from max U on escape path to find saddle point.	Static Push to escape. Minimize $\lvert\nabla U\rvert^2$ from max U on escape path to find saddle point.	Find sad. pt. and define S as $\perp$ hyperplane in subspace of participants, with bystanders clamped	MC Push using empirical reaction coordinate $r(q)$
Choose dividing Surface S	S = hyperplane $\perp$ to unstable normal mode	S = hyperplane $\perp$ to unstable normal mode		Define S by $r(q) = r_{min}$.
Sample S ensemble	Not necessary	Equilibrium MC run on or near the surface S	Equilibrium MC run on or near the surface S	Equilibrium MC run on or near the surface S
Calc. $Q^{\#}/Q_a$	Normal mode freqs. at s.p. and minimum	MC Calorimetry between A and S ensembles	MC Calorimetry between A and S ensembles	MC Calorimetry between A and S ensembles
Calc. $\langle\lvert u_\perp\rvert\cdot\hat{\xi}\rangle_s$		MD forw. & backward in time from (p,q) on S	MD forw. & backward in time from (p,q) on S	MD forw. & backward in time from (p,q) on S
Rate Const. (d)				

(Representative Trajectories through Bottleneck)

(a) U is quadratic within kT above and below local minima and saddle points.

(b) U is not quadratic, but is still 'smooth' on a scale of kT, so that adjacent local minima are typically separated by barriers higher than kT.

(c) U smooth with respect to some degrees of freedom, the 'participants', but rough on a scale of kT with respect to others, the 'bystanders'.

(d) Rate constant W is obtained by eq. 12. if U is harmonic, otherwise by eq. 7.

Acknowledgements

I wish to thank Phil Wolfe and Michael Levitt for valuable dicussions of minimization and escape methods, and Aneesur Rahman for repeatedly drawing my attention to practical cases of intolerable slowness in molecular dynamics. Some of the work was done at the Centre Européen de Calcul Atomique et Moléculaire, Orsay, France.

Literature Cited

1. Glasstone, S., Laidler, K.J., and Eyring, H., "Theory of Rate Processes" McGraw-Hill, New York, 1941
2. Alder, B. J., and Wainwright, T. E., J. Chem. Phys. (1959) 31, 459
3. Verlet, L., Phys. Rev. (1967) 159, 98
4. Rahman, A. and Stillinger, F.H., J. Chem. Phys. (1971) 55, 3336
5. Bunker, D.L., Methods Comp. Phys. (1971) 10, 287
6. Wood, W.W., and Erpenbeck, J.J., Ann. Rev. Phys. Chem. (1976) 27
7. Metropolis, N., Rosenbluth, A.W., Rosenbluth, M.N., Teller, A.H., and Teller, E., J, Chem. Phys. (1953) 21, 1087
8. Wood, W. W. in "Physics of Simple Liquids" (ed. H.N.V. Temperley, J. S. Rowlinson, and G.S. Rushbrooke) pp. 115-230, North-Holland, Amsterdam, 1968
9. Wood, W. W., in "Fundamental Problems in Statistical Mechanics III" (ed. E.G.D. Cohen) pp. 331-338, North-Holland, Amsterdam, 1974
10. Keck, J. C., Discuss. Faraday Soc. (1962) 33, 173
11. Anderson, J. B., J. Chem. Phys. (1973) 58, 4684
12. Wynblatt, P. J. Phys. Chem. Solids (1968) 29, 215
13. Jaffe, R.L., Henry, J.M. and Anderson, J.B, J. Chem. Phys. (1973) 59, 1128
14. Bennett, C. H., in "Diffusion in Solids: Recent Developments" (J. J. Burton and A. S. Nowick, ed.), pp. 73-113, Academic Press, New York, 1975
15. Bennett, C.H., 19'e Colloque de Métallurgie, Commissariat a l'Energie Atomique, Saclay, France 22-25 June 1976
16. Rahman, A., unpublished results
17. Bennett, C.H., J. Comp. Phys. (1976) 22, 245
18. Valleau, J.P., and Card, D.N. J. Chem. Phys. (1972) 57, 5457
19. Torrie, G. and Valleau, J.P., J. Comp. Phys. to be published
20. Pechukas, P. and McLafferty, F.J., J. Chem. Phys. (1973) 58, 1622

21. Chapman, S., Hornstein, S.M., and Miller, W.H., J. Am. Chem. Soc. (1975) 97, 892
22. Fletcher, R. Computer J. (1970) 13, 317
23. Fletcher, R. and Reeves, C.M., Computer J. (1964) 7, 149
24. McIver, J.W. and Komornicki, A., J. Amer. Chem. Soc. (1972) 94, 2625
25. Wigner, E., Z. Physik Chem. (1932) B19, 203
26. Johnston, H.S., "Gas Phase Reaction Rate Theory", pp. 133-134, Ronald Press, New York, 1966
27. ibid., pp. 190 ff.
28. Chapman, S., Garrett, B.C., and Miller, W.H., J. Chem. Phys. (1975) 63, 2710
29. Miller, W.H., J. Chem. Phys. (1974) 61, 1823
30. Johnston, op. cit., pp. 310 ff.
31. Levitt, M. and Warshel, A. Nature (1975) 253, 694
32. Levitt, M., J. Mol. Biol. (1976) 104, 59
33. Gibson, K.D. and Scheraga, H.A., Proc. Nat. Acad. Sci. USA (1969) 63, 9
34. Levitt, M. private communication
35. Bennett, C.H., Report of 1976 Workshop on Protein Dynamics, Centre Européen de Calcul Atomique et Moléculaire, Orsay 91405, France
36. Dewar, M.J.S. and Kirschner, S., J. Amer. Chem. Soc. (1971) 93, 4291
37. Bennett, C.H., J. Comp. Phys. (1975) 19, 267
38. Ryckaert, J.P., Ciccotti, G. and Berendsen, H.J.C., J. Comp. Phys., to be published
39. Torrie, G., Valleau, J.P., and Bain, A., J. Chem. Phys. (1973), 58, 5479

5

Newer Computing Techniques for Molecular Structure Studies by X-Ray Crystallography

DAVID J. DUCHAMP

The Upjohn Co., Kalamazoo, MI 49001

Crystallographers have been users of computers ever since computers became available for scientific calculations. The nature of crystallographic calculations used in molecular structure determination--large amounts of data to be treated by rather complicated mathematics--makes efficient use of computers essential and led quite early to the development of rather sophisticated techniques for both manual and computer computations. The features which make crystallographic calculations somewhat different include: 1) the use of symmetry, $i.e.$ space groups, 2) the use of a generalized coordinate system, 3) the three-dimensional nature of both data and intermediate and final results, 4) the high precision of the results, leading to generous use of statistics, 5) use of computer controlled data acquisition, and 6) the need for display and presentation of three-dimensional molecular structure information. For the most part, these are the areas in which crystallographers have tended to be in the forefront in algorithm development.

This paper concentrates on newer computing techniques, trying to give a sampling of recently developed techniques, which may be useful to both crystallographers and non-crystallographers. Material judged only understandable with in depth crystallographic background has been omitted. Apologies are made for the omission of many "favorite" algorithms. Since many of the algorithms are unpublished, the more detailed descriptions are taken of necessity from the author's own experience. The older algorithms not discussed here are well described in standard reference works, such as "The International Tables for X-ray Crystallography" (1) and textbooks by Rollett (2) and Stout and Jensen (3). In addition, many of the algorithms used in crystallographic computing are taken from numerical analysis (4) or are direct applications of standard computing algorithms such as those used in sorting data. The recent textbook of Aho, Hopcroft and Ullman (5) (and the references therein) provide an excellent introduction to the literature of general purpose computing algorithms, as well as an introduction to the strategies used in development of efficient algorithms.

Computing Techniques for X-ray Diffractometers

In most computer-controlled diffractometer systems, the computer has control of the settings and rate of change of the angles (usually 4) which determine the orientation of the crystal and the position of the radiation detector relative to the incident X-ray beam. It can also usually open and close the incident beam shutter, and control the counting of pulses from the detector. The basic process of data collection, which all systems can perform, consist of: for each reflection 1) calculate the settings of the angles, 2) move the diffractometer goniometer to those settings, 3) measure the intensity of the reflection, and 4) output the measured intensity. In addition most systems have enhancements, such as a program to aid in determining the orientation of the crystal on the instrument. Usually a fair amount of manual operation is required in setting up the experiment, including the correct indexing of the reflections.

In most cases, the crystallographer has little control over the computer programs, since they are most often coded in assembler language on a small minicomputer, and are therefore difficult to modify. In some laboratories, however, most of the programs are written in an easily changed high level language, making it easy to modify the algorithms used for programmed experiments, and to develop programs for new experiments. In the system in our laboratory (Figure 1), a small instrument control minicomputer operates as a slave to a larger lab automation computer. When a Fortran program running in the larger computer wants a specific task performed on the diffractometer, it loads a program into the minicomputer (unless the program is already there), and sends it information for the task to be performed. At task complete, the Fortran programs in the larger computer process the result and determine the course of the experiment. Getting a piece of information measured on the diffractometer is functionally similar to calling a subroutine which returns after the information is available. An alternative way to achieve the same flexibility is to build up the instrument control minicomputer into a much larger system.

Several improvements to the basic data collection algorithm have been made. Perhaps the most significant is the use of the step-scan technique, versions of which were developed in 1969 for our computerized diffractometer, and simultaneously elsewhere. The usual method of integrated intensity measurement is to scan continually through the reflection profile, accumulating counts continuously, then to measure the background by counting for fixed time at each extreme of the profile (6). Blessing, Coppens, and Becker have recently discussed the step-scan procedure (7). Basically it consists of sampling the peak profile at a number of points, perhaps 50 to 100, see Figure 2. Computer analysis of

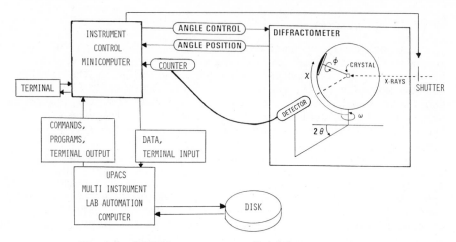

Figure 1. UPACS computer-controlled diffractometer system

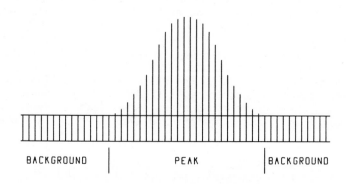

Figure 2. Step scan data collection

the recorded profile provides many advantages over the "blind"
continuous scan mode, allowing a much superior background correction, making possible the detection of abnormal profiles, and
producing a reduction in experimental standard deviations over
the former method. In addition the step-scan experiment is
generally faster since the time spent counting background is
eliminated. Further work on processing step-scan data ($\underline{8}$, $\underline{9}$)
and further work optimizing the measurement of x-ray intensities
($\underline{10}$, $\underline{11}$, $\underline{12}$) have recently appeared; the references in those
papers provide access to the earlier literature on this subject.

In addition to the improvement of the basic data collection
procedures, programs and algorithms are being developed for other
experiments to assist in the use of the diffractometer and to
make the process more automatic. Progress in this area has been
slow as recently pointed out by Spinrad ($\underline{13}$). The goal of being
able to drop a crystal in a magic funnel and have everything
happen automatically is not in sight, however, significant automatic enhancements are being made. Procedures to aid in indexing
reflections were developed by Sparks ($\underline{14}$) and more recently by
Jacobson ($\underline{15}$); in our laboratory a procedure involving somewhat
more interaction with the diffractometer is under development.
Two experiments which we have found very useful--precise alignment of the x-ray tube and determination of precision unit cell
parameters--are described in detail below.

When the x-ray tube is changed on a diffractometer it must
be positioned very precisely to center the x-ray beam in the
incident beam colimator. This is accomplished by translating the
tube in the plane perpendicular to the colimator. Approximate
positioning is easily accomplished manually. Then a test crystal
is placed on the diffractometer, and from angle values obtained
by centering certain reflections in the detector, misalignment of
the tube may be inferred. The process is complicated by slight
deviations of the crystal from the center of the goniometer (both
in height along the ϕ axis and translation (normal to it), the
arbitrary zero point of the ϕ angle, and possible misalignments
of the zero points of the 2θ, ω, and χ angles--all of which
affect the centering of a reflection in the detector. In our
procedure, the user mounts the test crystal, invokes the procedure and gives the computer approximate setting angles for one or
more reflections. The computer measures accurate centering
angles for each test reflection at the 8 possible positions with
$\omega = \theta$, as shown in Table I(a). From this data, a simple algorithm allows the computer to separate the different variables,
and to direct the user exactly (to within the approximation of
small translations) how far and in what direction to move the
tube, see Table I(b). Other valuable information derived from
this experiment are accurate determinations of the true zero's of
the ω, 2θ, and χ angles. The detailed equations are not presented here, since they vary with goniometer geometry, however a
short Fortran program for performing the calculation for the

Table I

a) Settings with $\omega = \theta$

2θ	ω	ϕ	χ
2θ	$2\theta/2$	ϕ	χ
-2θ	$-2\theta/2$	ϕ	χ
-2θ	$-2\theta/2$	ϕ	$\chi + 180$
2θ	$2\theta/2$	ϕ	$\chi + 180$
2θ	$2\theta/2$	$\phi + 180$	$-\chi$
-2θ	$-2\theta/2$	$\phi + 180$	$-\chi$
-2θ	$-2\theta/2$	$\phi + 180$	$180 - \chi$
2θ	$2\theta/2$	$\phi + 180$	$180 - \chi$

b) Computer report (retyped for clarity)

X-RAY ALIGNMENT REPORT AFTER-ADJUST-AGAIN 3/4/75   12812

H	K	L	TTH	OMEGA	PHI	CHI	INT
1	0	0	16.46	8.21	332.04	78.98	815
1	0	0	-16.45	-8.23	332.04	79.20	830
1	0	0	-16.44	-8.21	332.04	180+79.20	811
1	0	0	16.46	8.23	332.04	180+78.96	783
1	0	0	16.45	8.22	152.03	- 79.16	917
1	0	0	-16.46	-8.22	152.03	- 79.07	903
1	0	0	-16.46	-8.23	152.03	180-78.95	896
1	0	0	16.45	8.22	152.03	180-79.28	913

```
        PHI ERROR = -0.022    PHI (CORRECTED) = 332.062
        CHI (AVE) = 79.105    AVE DEL (CHI) = 0.110
            NEED TO MOVE TUBE DOWN 3.2 DIVISIONS
        CHI (ZERO) = -0.015
        OMEGA ERROR FROM CENTERING = -0.000
            PROBABLY CRYSTAL HEIGHT
        APPARENT TTH (ZERO) = 0.001
        APPARENT OMEGA (ZERO) = -0.000
            NEED TO MOVE TUBE OUT 0.3 DIVISIONS FOR TTH
            OR MOVE TUBE IN  0.1 DIVISIONS FOR OMEGA
```

Syntex diffractometer is available from the author on request.

Although a determination of the unit cell parameters results from determination of the orientation and indices of several reflections used to initiate the data collection experiment, we have found that a considerably more accurate determination may be made by running a separate experiment involving only measuring 2θ values for high 2θ reflections. Depending upon the crystal system, 1, 2, 4, or 6 of the unit-cell axial lengths and interaxial angles have to be measured experimentally, the remaining parameters being fixed by symmetry. The symmetry of the unit cell is important and must be used in precision unit-cell determination.

The procedure consists of four steps: 1) the computer surveying the intensities of previously measured reflections to choose about 20 high 2θ reflections, 2) making highly accurate step-scans of the selected reflections, 3) calculating accurate 2-theta values from the scan data; and 4) calculating unit-cell parameters from accurate 2-theta measurements.

The method used to calculate the "best" 2-theta for each reflection from step-scan data was developed especially for this system. Each peak is actually a doublet--one peak due to α_1 radiation and another due to α_2 radiation. The method assumes that this doublet may be fit by the sum of two Gaussian curves separated by $\Delta 2\theta$ which can be calculated from the wavelengths and the approximate 2-theta of the α_1 peak:

$$I_i = c\left[2e^{-\left(\frac{2\theta_i-2\theta_1}{w}\right)^2} + e^{-\left(\frac{2\theta_i-(2\theta_1+\Delta 2\theta)}{w}\right)^2}\right] + d \qquad (1)$$

where I_i is the calculated count at $2\theta_i$; w, c, and d are parameters dependent upon peak width, peak height, and background, respectively. The "best" 2-theta, $2\theta_1$ above, is calculated by a non-linear least-squares procedure which varies c, w, and $2\theta_1$ to minimize

$$\sum_{\substack{\text{all} \\ \text{steps}}} [g_i((I_i)_0 - (I_i)_c)]^2. \qquad (2)$$

where g_i is the weight calculated by taking the reciprocal of the standard deviation (from counting statistics) of $(I_i)_0$.

The value of d is calculated by averaging step-scan observations at ends of the scan, and is not varied during the least-squares procedure. Derivatives are calculated analytically using expressions obtained by differentiating equation 1. Up to 10 iterations are allowed; 3 to 5 are usually required.

When the method was developed, the effects of c, w, and d on $2\theta_1$ and $\sigma(2\theta_1)$, the error estimate for $2\theta_1$, were thoroughly

studied. The value used for d was found to have little or no
effect on either $2\theta_1$ or $\sigma(2\theta_1)$, unless an utterly ridiculous d
value was assumed. Therefore d is not in the refinement. The
values of c and w were found to have only small effects on $2\theta_1$,
but somewhat larger effects on $\sigma(2\theta_1)$. The two parameters c and
w are strongly correlated--allowing large shifts in c before w
has quieted down results in an unstable refinement.

Calculation of the unit-cell parameters from the 2θ data is
accomplished by a special adaptation of a method used in several
laboratories for determining accurate cell parameters from special film data (16). For the general case

$$\frac{4}{\lambda^2} \sin^2\theta = h^2 a^{*2} + k^2 b^{*2} + \ell^2 c^{*2} + 2k\ell b^* c^* \cos\alpha^* +$$

$$2h\ell a^* c^* \cos\beta^* + 2hk \, a^* b^* \cos\gamma^* \tag{3}$$

where h, k, and ℓ are reflection indices; a^*, b^*, c^*, α^*, β^*, γ^*
are axial lengths and interaxial angles of the reciprocal cell.
Equation 9 may be abbreviated as

$$\sin^2\theta = h^2 s_1 + k^2 s_2 + \ell^2 s_3 + k\ell s_4 + h\ell s_5 + hk s_6 \tag{4}$$

The linear least-squares procedure determines $s_1, \ldots, s_6$
so as to minimize

$$\sum_{i=1}^{N} w_i^2 [(\sin^2\theta_i)_0 - (\sin^2\theta_i)_c]^2 \tag{5}$$

Comparison of equations 3 and 4 shows immediately how to
calculate the reciprocal cell parameters from the coefficients in
4. From these, the unit cell parameters may be calculated using
standard expressions (17). The weight of each observation is
calculated by

$$w_i = \frac{1}{(\sin 2\theta)\sigma(2\theta)} \tag{6}$$

The effect of symmetry is conveniently taken into account by
restrictions on $s_1, \ldots, s_6$ as follows:

Crystal System	To Be Determined	Restrictions
Triclinic	$s_1, \ldots, s_6$	None
Monoclinic	s_1, s_2, s_3, s_5	$s_4 = s_6 = 0$
Orthorhombic	s_1, s_2, s_3	$s_4 = s_5 = s_6 = 0$
Tetragonal	s_1, s_3	$s_2 = s_1, \quad s_4 = s_5 = s_6 = 0$
Hexagonal	s_1, s_3	$s_2 = s_6 = s_1, \quad s_5 = s_4 = 0$
Cubic	s_1	$s_3 = s_2 = s_1, \quad s_4 = s_5 = s_6 = 0$

Standard deviations in unit-cell parameters may be calculated analytically by error propagation. In these programs, however, the Jacobian of the transformation from s_1, ... , s_6 to unit-cell parameters and volume is evaluated numerically and used to transform the variance-covariance matrix of s_1, ... , s_6 into the variances of the cell parameters and volume from which standard deviations are calculated. If suitable standard deviations are not obtained for certain of the unit cell parameters, it is easy to program the computer to measure additional reflections which strongly correlate with the desired parameters, and repeat the final calculations with this additional data.

Treatment of Crystal Deterioration:

The variation of the integrated intensities of X-ray reflection as a function of time of exposure to X-rays is a problem which has plagued crystallographers for some time. Little is known of the physical and chemical processes leading to radiation damage (<u>18</u>). Usually several carefully chosen reflections (check reflections) are repeated at regular intervals during data collection. The problem is how best to use the fluctuations in these measured intensities to scale the observed set of intensities. We use 10 check reflections after experimenting with more and fewer. Since the fluctuations of intensity with time are almost always non-linear, and frequently are non-monotonic also, a fairly complicated function is required to express the deterioration scale factor.
In the procedure described here, the scale factor is represented as a function of time $C(t)$ described mathematically by

$$C(t) = a_1 f_1(t) + a_2 f_2(t) + ... + a_p f_p(t) \qquad (7)$$

where t is the cumulative exposure time of the crystal, the $f_k(t)$ are functions of t, and the a_k are the coefficients to be determined from the check reflection data to specify $C(t)$. The criteria chosen is to determine the a_k so as to minimize the sum of the weighted second moments about the means of the scaled check reflection intensities. With a second Lagrange undetermined multiplier term added to avoid the trivial minimum, the function minimized becomes

$$\sum_{i,j} w_j^2 (b_j - C(t_{ij})g_{ij})^2 - \sum_{i,j} C(t_{ij})^2 \qquad (8)$$

where t_{ij} is the time for the ith observation of the jth check reflection, g_{ij} is its intensity. The weights w_j are defined by

$$w_j = m_j^{-1} \sum_i \sigma_{ij}^{-1} \tag{8}$$

where σ_{ij} is the standard deviation in g_{ij}, and m_j is the number of observations of check reflection j. The b_j are defined by

$$b_j = m_j^{-1} \sum_i C(t_{ij})g_{ij} \tag{9}$$

By suitable mathematical manipulation the above may be shown to be a linear least-squares with constraint problem in the variables a_k. Before the a_k can be determined, the functions $f_k(t)$ must be specified.

If $C(t)$ is chosen to be a simple polynomial in t, (*i.e.*, $f_k(t) = t^{k-1}$), and a direct least-squares solution is calculated, calculation trouble usually results since the determinant of the coefficients of the normal equations tends to be very small (<u>19</u>). A $C(t)$ with all the flexibility of the general polynomial is obtained, and the numerical problem is avoided by choosing the $f_k(t)$ to be the orthogonal polynomials of Forsythe (<u>19</u>). Cast in our notation, the $f_k(t)$ are defined recursively by

$$f_1(t) = 1$$

$$f_2(t) = (t - u_2)f_1(t)$$

$$f_3(t) = (t - u_3)f_2(t) - v_2f_1(t)$$

$$\vdots$$

$$f_k(t) = (t - u_k)f_{k-1}(t) - v_{k-1}f_{k-2}(t) \tag{11}$$

where

$$u_k = \frac{\sum_{i,j} t_{ij}(f_{k-1}(t_{ij}))}{d_{k-1}} \tag{12}$$

$$v_{k-1} = \frac{\sum_{i,j} t_{ij}f_{k-1}(t_{ij})f_{k-2}(t_{ij})}{d_{k-2}} \tag{13}$$

$$d_k = \sum_{i,j} (f_k(t_{ij}))^2 \tag{14}$$

In this formulation, the needed coefficients a_k may be calculated directly without recourse to solving the usual eigenvector problem.

In our programs provision is also made for a dependence of scale factor on direction in the crystal, $\underline{h}$, and on the Bragg angle, θ. A new scale factor $C'(t,\underline{h},\theta)$ is defined as

$$C'(t,\underline{h},\theta) = 1 + (C(t)-1) \; H(\underline{h}) \; E(\theta) \qquad (15)$$

where $C(t)$ is our original function in time, $H(\underline{h})$ is a direction dependent factor with six determinable parameters, and $E(\theta)$ is a factor with one determinable parameter. The coefficients of this generalized scale factor function is determined to minimize the same quantity with C' replacing C, by first solving as before with the new parameters set so that $H(\underline{h}) = E(\theta) = 1.0$, then allowing all parameters to vary from that point in an iterative minimization procedure similar to "steepest descents". A more detailed description of the generalized scale factor function is contained in an implementation of this scaling algorithm in a Fortran data reduction program available from the author.

Hidden Line Algorithms:

In the display of a three dimensional object on a plotter or on the screen of a graphics terminal, the task of deciding which parts of the object should be shown and which should be eliminated (or made dashed) is known as the "hidden line problem". This problem and the more complicated "hidden surface problem" has recently been reviewed by Sutherland, Sproull and Schumacker (20) from a sorting point of view. These algorithms are especially important because programs with inefficient hidden line algorithms can use up enormous amounts of computer time and because manual "touch up" of drawings to eliminate hidden line errors may be quite time consuming. The most efficient algorithms result when the object to be drawn has special features which allow the general problems to be simplified. Two problems are treated here in some detail: the drawing of a crystal from face measurements and the drawing of a "ball and stick" representation of a molecule.

The problem of producing of a crystal like that shown in Figure 3 arose in a graphics program (21) used to visually compare the computer description of a crystal as a convex polyhedron with the crystal as viewed on an optical goniometer. The problem is one of displaying a convex polyhedron given the information describing the faces of the polyhedron. From this information the faces which intersect at the various corners and the coordinates of the corners can easily be computed (22). From this, a list of edges--the lines actually to be drawn in the figure--can easily be compiled.

In producing the drawing, a rotation of the coordinates of the corners is performed to give a set of x,y,z relative to an origin at the center with the x axis aligned with the viewing direction. Next is identification of those edges which lie on the convex polygon which defines the periphery of the polyhedron in projection on the y,z plane. For each edge, defined by two corners i and j, the edge is on the polygon if all other corners either lie on the edge or on one side of it in projection on the y,z plane, or simply if

$$(z_i-z_j)y_k + (y_j-y_i)z_k + y_iz_j - z_iy_j \quad \begin{array}{l} \geq 0 \text{ for all } k \\ \text{or} \\ \leq 0 \text{ for all } k \end{array} \quad (16)$$

For simplicity in practice, the =0 case is eliminated by slight translation of corner coordinates. All other edges are either "totally hidden" or "totally visible". The "hidden line" problem, therefore, becomes one of classifying the edges (the lines actually drawn) into one of the three categories. Also a "totally hidden" edge may not connect with a "totally visible" edge except through one of the corners on the peripheral polygon. Because of the convex property of the polyhedron, other edges may be classified by connectivity if one edge not on the polygon is classified. This is accomplished easily by finding two edges defined by corners i,k and i,j where corners i and j are on the polygon and k is not. The edge defined by corners i,k is either "totally visible" or "totally hidden" according as a and d defined below have the same or opposite signs, respectively.

$$a = y_k(z_i-z_j) + y_i(z_j-z_k) + y_j(z_k-z_i) \quad (17)$$

$$d = x_k(y_iz_j-y_jz_i) + x_i(y_jz_k-y_kz_j) + x_j(y_kz_i-y_iz_k) \quad (18)$$

As many unclassified edges are classified by connectivity as possible. Then if unclassified edges remain, equations 17 and 18 are used to classify another, etc., until all edges are classified.

In the DRAW program which we developed, the "hidden line" algorithm for ball and stick drawings of molecules (such as Figure 4) likewise makes use of special features of the object. The drawing is composed of only two kinds of figures--circular atoms and trapezoidal bonds. Our algorithm is similar to one developed by Okaya (23). The more complicated case of general elipsoidal representation of atoms has been treated by Johnson in the latest version of his heavily used ORTEP program (24).

In principle, when each atom or bond is drawn, it must be tested against all other bonds and atoms to see if it is hidden, totally or in part. In the drawing operation, each atom or bond is represented by a number (usually 100 to 200) of points with

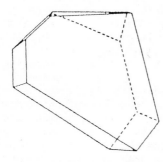

Figure 3.   Computer drawing of crystal from face description

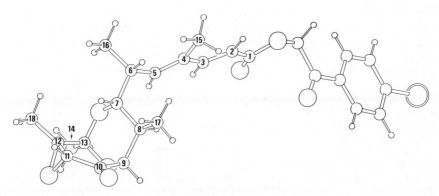

Figure 4.   Ball and stick drawing of molecule of p-bromophenacyl ester of tirandamycic acid

straight lines connecting them; a separate visibility test must
be made on each point in deciding whether to draw the lines to
and from it. As the number of atoms (n) grows, the complexity of
the calculation increases as n^2. By using an application of the
"divide and conquer" strategy (25), the problem is reduced to a
very quick approximately n^2 complexity part and a more time
consuming almost n complexity part. At the time each atom or
bond figure is drawn, a quick test is employed to compile a list
of those atoms or bonds which could possibly overlap in the
figure. In practice the size of this list, after reaching a
certain level, does not increase very much as n increases. This
is easily understood by considering that: 1) for m randomly
distributed objects within a volume, the "object thickness" is
the cube roote of m, and 2) in order to make a drawing understandable, people usually draw figures with minimum overlap in
the projection direction. Therefore the time consuming point by
point tests are performed only on a greatly reduced number of
figures. A number of enhancements can be made to the point-by-
point test which speed it up but do not reduce its complexity.
On the other hand, in principle, the complexity of the pretest
portion can be reduced from n^2 to $n^{3/2}$ by ordering the bonds and
atoms in the longest direction in the plane of projection, and
only testing figures lying in a relevant band. Since the pretest
is so fast, we have not implemented this final refinement in the
batch versions of our program; however, it is under consideration
for a graphics version now being implemented.

Use of the Fast Fourier Transform:

Although the principle of the fast Fourier transform (FFT)
algorithm has been widely understood for over ten years (26, 27),
the FFT is only now beginning to be used widely for crystallographic calculations. The reasons for this are: 1) the
advantages of the FFT are not nearly as great in crystallographic
computing as in other fields, 2) crystallographic trigonometric
Fourier algorithms (28) have been highly developed and are very
efficient, and 3) incorporation of the special features of
crystallographic calculations, such as symmetry, has required
additional algorithm development.

In the simpler FFT applications to chemistry, such as in
Fourier transform spectroscopy, the tremendous advantage of the
FFT algorithm arises because for computing n Fourier coefficients
from n data points, the FFT algorithm reduces the complexity from
n^2 to n log n. This is brought about by factoring the transform
very finely so as to allow calculations common to several transformed points to be performed only once. In the efficient formulations of the crystallographic trigonometric Fourier algorithm, a certain amount of factoring is employed, leading to a
complexity of approximately $n^{4/3}$ (29) instead of the usually
quoted n^2. In an early comparison (29), factors in improvement

by use of the FFT of 1.8 to 19.0 were achieved by use of the FFT
algorithm; verifying that the thousand fold gains found in other
areas are not present in the crystallographic case. For high n a
point is reached where the FFT is more efficient. The size of
the problem necessary for the FFT algorithm to be considerably
faster depends on the efficiency of the implementations of the
respective algorithms; *i.e.*, it depends upon the coefficients
which multiply the complexity factor to give the cost of the
calculation. It is not surprising that the area which is making
the most use of the FFT is macromolecular crystallography where
values of n are usually very large.

Considerable work has been done recently on the problems of
developing the FFT for crystallographic use. The problem of
incorporating space group symmetry has been elegantly treated by
Ten Eyck (30) and in a simpler fashion by Bantz and Zwick (31).
Other implementations include those of Immirzi (32) and Lange,
Stolle and Huttner (33), both of which treat the problem of the
enormous amount of computer storage required to store an entire
crystallographic map (100,000 to 500,000 points are frequently
required), and also the work of Mallinson and Teskey (34), which
discusses the problem of handling negative indices economically.

In the future the FFT algorithm will be more widely used in
small molecule as well as macromolecular crystallography, especially as new efficient FFT programs are integrated into the
various program systems used for such calculations. In practice,
a good general purpose program (especially efficient for small
molecule crystallography) could be developed by combining the
strengths of the FFT and trigonometric techniques. The crystallographic Fourier transform, whether it be done by FFT or other,
can be factored into three parts, a "first dimension" in which
summation is made over the direction normal to the sections of
the three-dimensional map, and a second and third dimension in
the plane of the map sections. A computer formulation of the
trigonometric triple product technique which incorporates the
space group symmetry almost exclusively in the first dimension of
the calculation is available (35). A program which performs the
first dimension calculation in the traditional space-group
specific manner, and performed the second and third dimensions by
the FFT algorithm would have several advantages. It would make
efficient use of the fact that in most crystallographic Fourier
calculations there are 10 to 20 times more calculated grid points
than input data, without having to resort to less efficient
formulations of the FFT algorithms which require complex multiplication. It would greatly alleviate the storage problem, and
would remove most of the symmetry considerations from the FFT
portion of the calculation, leading to a simpler implementation
at the inner most part of the calculation. This proposed program
bears some similarity to the work of Immirzi (32), where the FFT
was not used in the first dimension because of storage considerations, but where symmetry was avoided by transforming the data

to triclinic. In the limit of high n, the proposed program would
of necessity be slower than an all FFT program. In the case of
small molecule E-maps, where the ratio of grid points to data is
especially high, this program would be most efficient, if done
right, considerably more efficient than an all FFT implementation.

Direct Methods:

Direct methods is the most widely used techinque for getting
a trial structure in small molecule crystallography, and has
increasing applications in macromolecular crystallography as well
($\underline{36}$). The problem is one of finding a set of approximate phases
$\phi_{\underline{h}}$ to assign the observed normalized structure factor magnitudes

$|E_{\underline{h}}|$ so that a Fourier transform calculation can be performed to
give an electron density map from which atomic positions can be
derived. Most computer programs for direct methods are based on
the Σ_2 formula ($\underline{37}$, $\underline{38}$) and the tangent formula ($\underline{38}$), both of
which relate phases by equations which have calculated proba-
bilities of being correct. The phases related in both cases are
those of reflection triples for which

$$\underline{h} + \underline{k} + \underline{\ell} = 0 \qquad\qquad (19)$$

where $\underline{h}$, $\underline{k}$, and $\underline{\ell}$ are vectors whose components are the integer
indices of the reflections which have large $|E|$. The algorithm
used to search for these triples is of primary importance to the
efficiency of most direct methods computer programs. The set of
high $|E|$ reflections usually comprise 0.1 to 0.3 of the symmetry
independent reflections. In the search, all the symmetry related
reflections must be used for two of the reflections; in ortho-
rhombic, for example, the symmetry independent set must be
expanded 8-fold either prior to the calculation or during each
test. The obvious three-loop way of finding triples leads to a
n^3 complexity algorithm (and a lot of wasted computer time).
This can be changed to an n^2 complexity procedure if each reflection is associated uniquely with an array subscript by some
equation involving the integer indices, so that given $\underline{h}$ and $\underline{k}$,
the subscript of $\underline{\ell}$ can be calculated and the presence of $E_{\underline{\ell}}$
in the set can be check by table lookup.

Perhaps the most efficient algorithm (used in several
programs, including the program DIREC written by the author) is
one originally developed by Dewar for the MAGIC program ($\underline{39}$).
Prior to the searching operation, the set of high $|E|$ reflections
is expanded to the full set of reflections, and the $\underline{h}$ vectors are
transformed into a set of real integers $\{m_i\}$ in such a way as to
preserve the arithmetic relationship among the $\underline{h}$. One such
mapping is

$$m = 1000000\, h_1 + 1000\, h_2 + h_3 \qquad\qquad (20)$$

where $\underline{h}$ vector components are h_1, h_2, h_3. Since the range of
possible values of h_1, h_2, and h_3 is restricted, if equation (19)
holds, the m values derived from the three vectors will also sum
to zero, and vice versa. Next the m_i are sorted numerically with
elimination of duplicates from the symmetry expansion. During
these operations a pointer back to the original reflection and a
symmetry operation code are carried along with each m_i. The
process of finding all $\underline{k}$ and $\underline{\ell}$ which form triples with $\underline{h}$, is thus
transformed to the problem of finding all pairs of integers from
the ordered set $\{m_i\}$ which sum to -n, where n is the "m value" of
$\underline{h}$. The transformed problem has a very efficient solution involving only one pass through $\{m_i\}$. Two pointers (i and j) are
initialized to point at the beginning and at the end of the set
respectively; all triples are found by moving i and j toward each
other until they meet, using the procedure diagrammed in Figure
5. The simplicity of this procedure can readily be appreciated
if the reader will construct an ordered array of 10 to 15 integers (in the range -20 to 20), and follow the algorithm to find
pairs which sum to a given value. Alternatively the pointers
could be started at n as favored by Dewar (39), and moved outward
in a linear sweep using a similar procedure.

The algorithm described above for finding triples may be
extended to find higher order relationships, for example, the
quartets (four vectors, $\underline{h}$, $\underline{k}$, $\underline{\ell}$, and $\underline{m}$ sum to zero) for which new
powerful formulas are being developed by Hauptman (40). However,
simple extension of this algorithm does not appear to be optimal,
and more research in this area is needed.

When the phase relationships and their probability have been
derived, several thousand inconsistent equations in a few hundred
unknowns must be translated into a set (or sets) of phases. The
procedures used for this are very interesting, but too specific
to crystallography to be discussed in detail here. One or more
specially chosen phases (depending on the space group) may be
assigned "free" to fix the degrees of freedom. Next the set of
known phases usually is extended by: 1) symbolic addition (41),
wherein symbols of unknown value are assigned to a few selected
reflections, and the set is extended by algebraic manipulations
which assign phases as linear combinations of symbols; or 2) the
multi-solution method (42) wherein all combinations of possible
phase values for a few reflections are carried through the extension to give a number of possible phase sets. The next step
is to rank the phase sets which result from the multi-solution
method, or from the assignment of numeric phases to the symbols
used in the symbolic addition method; no foolproof way to do this
has yet been found. Frequently several, sometimes many, sets of
phases must be tried before a trial structure is obtained. With
enough perseverance, however, a trial structure can almost always
be obtained by direct methods using presently available programs.
New theoretical developments in direct methods hold promise for
improved, more automatic computer programs for determining

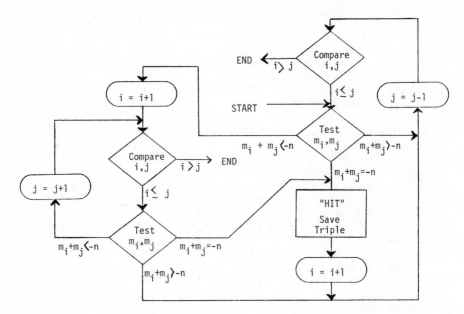

Figure 5.   Procedure for finding all pairs of integers with a given sum

starting phase sets.

Molecular Mechanics "Strain Energy" Calculations:

Since molecular mechanics "strain energy" calculations (43, 44) have become a valuable tool in interpretation of molecular structure results from crystallographic studies, certain computing techniques used there will be mentioned. The method is simple in principle; the strain energy of a particular conformation of a molecule is expressed as the sum of terms of several types, each related to certain structural parameters; for example, bond length, non-bonded contacts, torsion angle.

$$E_{strain} = E_{bond} + E_{angle} + E_{torsion} + \ldots \qquad (21)$$

Each term is a simple equation involving one or more empirically derived potential parameters and one or more structural parameters. In the usual calculation, the structural parameters are varied to minimize the strain energy, the potential parameters being held fixed. Crystal structure results are sometimes used to derive potential parameters (45, 46).

In most studies of molecular structure starting from crystallographic results, it is useful to calculate the minimum energy for the molecule in the crystal. Usually the molecule may be surrounded by its nearest neighbors in the crystal, and the minimization may be carried out by holding the unit cell parameters fixed and varying the atomic positions, with preservation of space group symmetry. This simple method will produce good results (provided suitable potential parameters are used) if calculation of the minimum energy molecular conformation is desired. It will not suffice if either the unit cell parameters are to be varied, intermolecular potential parameters are to be varied, or if accurate lattice energies are to be calculated. For these purposes lattice sums should be evaluated; a particularly efficient method for doing this is the convergence acceleration algorithm of Williams (47).

In our experience, the introduction of "extra potentials" is a particularly useful technique when molecular conformations other than the minimum energy one must be explored. In this method, potentials are added which make it prohibitively expensive (in energy terms) for the molecule not to assume the desired structural feature. The total energy--strain energy plus "extra potential" energy--is minimized, giving the minimum energy conformation of the molecule subject to the constraint imposed by the "extra potentials".

$$E_{total} = E_{strain} + \sum E_{extra} \qquad (22)$$

By subtracting the strain energy portion of the total energy from
the strain energy of the molecule in its minimum energy conformation, the cost of assuming the non-minimal conformation may be
assessed. Many properties of molecules may be conveniently
studied by this technique, including: flexibility of the molecule
with respect to a certain torsion angle, barriers between conformational minimas, and the feasibility of certain conformations
predicted to be "active". One application we have found especially useful is the matching of two molecules which are presumed
to bind at the same active site. In this procedure, (see Figure
6) two or more molecules are minimized simultaneously while being
linked at certain selected sites by "extra potentials". One word
of caution is appropriate here--"extra potentials" are usually
set to be so strong that, without due care, the calculation may
become unbalanced, causing certain minimization techniques to
converge quite slowly. This is particularly true of certain
"pattern search" routines (48) used in many programs.

Minimization techniques are of great importance to both the
efficiency of molecular mechanics computer programs, and the
accuracy and reproducibility of the results. The energy expression is non-linear in the variables used in the calculation. If,
as is usual, atomic coordinates are the variables, the number of
variables is greater than the number of degrees of freedom. The
energy surface is characterized by many local minima; and by the
fact that a minimum is frequently quite flat for considerable
distances in parameter space. An optimal minimization algorithm
for such problems is yet to be discovered. Methods currently
used include search techniques, which converge from large distances, but are inefficient in flat minima, and more complicated
methods such as Newton's Method, which works well in finding the
minimum but is extremely time consuming if the initial starting
point is far off.

Automating Crystallographic Calculations:

During the course of a crystal structure determination a
large number of different types of calculations must be performed. Prior to the advent of crystallographic computing systems,
each type was incorporated into a different program with its own
peculiar form of input and output. With the advent of programming systems (49, 50) must of the incompatibility between programs, and much of the tedium of crystallographic computing, was
eliminated--how much so depends upon the particular system.

A reasonable set of goals to strive for in automating a
computing process are:

a)   single entry of data
b)   minimization of input, including providing defaults for
      all options and not requiring entry of anything the
      computer can calculate

c)   minimization of input errors
d)   computer runs until a "human" decision is needed
e)   minimum effort for a decision
f)   minimum effort to implement decisions

For example, if the crystallographic unit cell parameters are
entered during data reduction calculations, they should not have
to be entered again in any subsequent calculation. One of the
best examples of not minimizing input is a computer program which
requires the user to enter the number of atoms to be entered,
instead of counting the atoms as they are entered. Good input
engineering, including the use of alphabetic labels and free
format where appropriate, will minimize input errors. Factors
which can minimize decision making effort include: organization
of data pertinent to the decision in a short summary form, and
presenting it in a way it can be quickly assimilated by the user.
After the requisite decisions are made, we can't say to the
computer "continue with calculation x", but we should strive to
come as close as possible to this.
   There are problems complicating this automation process,
some are computer engineering, some practical, and some basically
philosophical. These include: the necessity for retaining
optional ways of doing the calculations, the need for the user to
retain control of the process, the restrictions placed on operation by the various computer systems, avoiding the waste of
computer time, and the inherent difficulty usually encountered in
automating decision making.
   A certain level of automation of the decision making and
decision implementation processes has been achieved in our
laboratory through use of a graphics terminal on-line to our
large research computer (21). Figure 7 shows the operational
hookup. Our graphics programs run in a high priority partition
in what is essentially a batch processing system. On-line disk
libraries are used to pass data between our graphics programs and
our regular batch calculations which run at a lower priority.
All our batch jobs are submitted through the graphics terminal,
including the job which transfers the initial data from the
laboratory automation computer to the large computer. Any time
consuming calculations are run in batch mode. For example,
electron density maps are calculated in a batch run, with the
results being saved in a disk library; a graphics program is used
for interpretation of the map since "human" decision is usually
required. The use of this graphics terminal has cut the amount
of people time required to run a series of crystallographic
calculations by more than a factor of two.
   In the area of input engineering, in the current version of
the CRYM system (developed by the author), excluding the job
control, 15 input records (card images) are required in one batch
run to take an initial set of data through a variety of data
reduction calculations, approxiate scaling, a direct methods

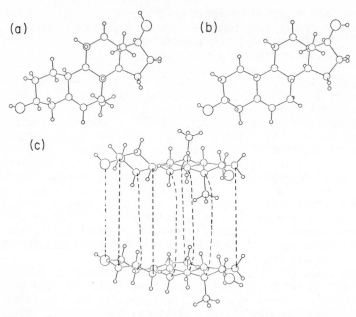

(a)   (b)

(c)

Figure 6.   Two Steroids—19-nor androstenediol (a); and 7-αme-19-nor androstenediol (b)—as found in crystal (viewed normal to C ring). Dotted lines in (c) show a possible placement of "extra potentials" for linking the molecules during simultaneous strain energy minimization.

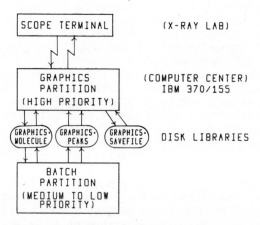

Figure 7.   Graphics system for crystallographic computing at Upjohn

calculation, and calculation of the most probable E-map ready for interpretation on the graphics terminal. Analysis of this input shows that for the case of the morphine free base structure it could be reduced to the four records shown below for the computation of the 4 most probable maps.

DATA REDUCTION (MORPHINE), SPACE GROUP = 19

ASYMMETRIC UNIT C17 H21 O4 N

DIRECT METHODS

EMAP, 1-4, (MORPEMC*)

By use of suitable abbreviations, a shorter form is possible.

DR(MORPHINE),SG=19

AU C17 H21 O4 N

DM

EM,1-4,(MORPEMC*)

Our system does not have this type of input, but it illustrates the direction we are headed. It is a worthwhile direction for any system of programs with a long lifetime.

ABSTRACT

This review presents a selection of newer algorithms used in X-ray crystallographic calculations. Some of the material is not previously published. Areas discussed in detail include: Algorithm design for computer-controlled diffractometers, a scheme for computer-aided alignment of X-ray tubes, a procedure for determining precision unit cell parameters, a method for scaling intensity data for crystal deterioration, "hidden line" algorithms for drawing crystals from face descriptions and for drawing ball and stick molecules, crystallographic use of the "fast Fourier transform" method, use of "extra potentials" in molecular mechanics, and the total automation of the X-ray computing process.

Literature Cited

1. "International Tables for X-ray Crystallography", Vol. I, II, III, IV, Kynoch Press, Birmingham.
2. Rollett, J. S., "Computing Methods in Crystallography", Pergamon Press, Oxford (1965).

3.   Stout, G. H. and Jensen, L. H., "X-ray Structure Determination", Macmillan, New York (1968).
4.   Abramowitz, M. and Stegun, I. A., Editors, "Handbook of Mathematical Functions", National Bureau of Standards, Government Printing Office, Washington (1964).
5.   Aho, A. V., Hopcroft, J. E., and Ullman, J. D., "The Design and Analysis of Computer Algorithms", Addison-Wesley, Reading, Massachusetts, (1974).
6.   Furnas, T. C., Jr., "Single Crystal Orienter Instruction Manual", General Electric Company, X-ray Department, Milwaukee, (1957).
7.   Blessing, R. H., Coppens, P., and Becker, P., *J. Applied Crystallography*, (1972), 7, 488.
8.   Lehmann, M. S. and Larsen, F. K., *Acta Cryst*, (1974), A29, 216.
9.   Lehmann, M. S., *J. Applied Crystallography*, (1975), 8, 619.
10. Mackenzie, J. K. and Williams, E. J., *Acta Cryst*, (1973), A29, 201.
11. Killean, R. C. G., *Acta Cryst*, (1973), A29, 216.
12. Grant, D. F., *Acta Cryst*, (1973), A29, 217.
13. Spinrad, R. J., "Abstracts of the American Crystallographic Association", Clemson, Series 2, 4, 35, (1976).
14. Sparks, R. A., "Abstracts of the American Crystallographic Association", Ottawa, (1970).
15. Jacobson, R. A., *J. Applied Crystallography*, (1976), 9, 115.
16. Original reference unknown; the author first encountered this method in a class on advanced x-ray crystallography taught by R. E. Marsh at Caltech.
17. Ref. 1, Vol. II, p. 106.
18. Abrahams, S. C., *Acta Cryst*, (1973), A29, 111.
19. See for example, L. G. Kelly, "Handbook of Numerical Methods and Applications", p. 66, Addison-Wesley, Reading, Massachusetts (1967), and the references therein.
20. Sutherland, I. E., Sproull, R. F., and Schumacker, R. A., *Computing Surveys*, (1974), 6, 1.
21. Duchamp, D. J., "Abstracts of American Crystallographic Association", Clemson, Series 2, 4, 20, (1976).
22. Busing, W. R. and Levy, H. A., *Acta Cryst*, (1957), 10, 180.
23. Okaya, Y., IBM Research Report, R.C. 1706, IBM Watson Research Center, Yorktown Heights, N.Y., (1966).
24. Johnson, C. K., ORTEP, ORNL-3794, Oak Ridge National Laboratory, Oak Ridge, Tennessee (1965).
25. Reference 5, p. 60.
26. Cooley, J. W. and Tukey, J. W., *Math. Comput.*, 19, 297.
27. Gentleman, W. M. and Sande, G., *Proceedings of the Fall Joint Computer Conference*, (1966), 563.
28. See reference 1, Vol. II, p. 78, for example.
29. Hubbard, C. R., Quicksall, C. O., and Jacobson, R. A., *J. Applied Crystallography*, (1972), 5, 234.
30. Ten Eyck, L. F., *Acta Cryst*, (1973), A29, 183.

31. Bantz, D. A. and Zwick, M., *Acta Cryst*, (1974), A30, 257.
32. Immirzi, A., *J. Applied Crystallography*, (1973), 6, 246.
33. Lange, S., Stolle, U., and Huttner, G., *Acta Cryst*, (1973), A29, 445.
34. Mallinson, P. R. and Teskey, F. N., *Acta Cryst*, (1974), A30, 601.
35. Duchamp, D. J., Thesis, California Institute of Technology, p. 82 (1965).
36. Sayre, D., *Acta Cryst.*, (1974), A30, 180.
37. Karle, J. and Hauptman, H., "Solution of the Phase Problem. I. The Centrosymmetric Case", Am. Cryst. Assoc. Monograph No. 3, (1953).
38. Karle, J. and Hauptman, H., *Acta Cryst*, (1958), 11, 264.
39. Couter, C. L. and Dewar, R. B. K., *Acta Cryst*, (1971), B27, 1730.
40. Hauptman, H., *Acta Cryst*, (1975), A31, 680.
41. Karle, J and Karle, I. L., *Acta Cryst*, (1966), 21, 849.
42. Germain, G., Main, P., and Woolfson, M. M., *Acta Cryst*, (1970), B26, 274.
43. Kitaigorodsky, A. I., "Molecular Crystals and Molecules", Chapter VII, Academic Press, New York (1973).
44. Engler, E. M., Andose, J. D., and von R Schleyer, P., *JACS*, (1973), 95, 8005.
45. Williams, D. E., *Acta Cryst*, (1974), A30, 71.
46. Ermer, O. and Lifson, S., *JACS*, (1973), 95, 4121.
47. Williams, D. E., *Trans. Amer. Cryst. Assoc.*, (1970), 6, 21.
48. See for example, R. Hooke and T. A. Jeeves, *J. Assoc. Computing Machinery*, (1961), 8, 212.
49. Duchamp, D. J., "Abstracts American Crystallographic Association", Bozeman, Montana, 29 (1964).
50. Stewart, J. M. and High, D. F., *ibid.*

6

Algorithms in the Computer Handling of Chemical Information

LOUIS J. O'KORN

Systems Development Dept., Chemical Abstracts Service,
The Ohio State University, Columbus, OH 43210

The chemical literature emphasizes the detailed structural characteristics of chemical substances; this paper addresses computer-based algorithms that support the handling of information about chemical substances. The nature of problems requiring an algorithmic solution, examples of specific algorithms to support these solutions, and some of the continuing problems are discussed. Since representation affects the nature of algorithms, several of the computer representations of a chemical substance are mentioned. For these representations, algorithm developments that perform interconversion, registration, and structure searching are discussed.

Introduction

The techniques utilized in chemical information handling systems fall into two categories -- those which handle the processing of text and those concerned with the processing of chemical substance information. The general text handling processes in chemical information handling systems are not substantially different from the processes of information handling systems for other scientific disciplines.

Although not discussed here, substantial development has occurred in the development of computer-based algorithms for text information handling systems. These computer-based text information handling systems provide for data base compilation to support traditional printed publication and also the selective dissemination of the information.

Algorithm development in the areas of computer editing, data base management, sorting, computer-based composition, and text searching have been critical to the overall development of computer-based primary and secondary publications systems and text search services. Results of these developments are illustrated in the computer-based information system used at Chemical Abstracts Service (CAS) [1]. Lynch [2] describes principles and techniques for the computer-based information services and

Cuadra [3] provides annual reviews of developments in information handling.

It is the set of methods for representing, sorting, manipulating and retrieving information about chemical substances that distinguishes the techniques of chemical information handling from those of other disciplines. Chemical literature emphasizes the detailed structural characteristics of chemical substances. This is illustrated by the fact that for the 392,000 documents abstracted in 1975 in CHEMICAL ABSTRACTS, 1,514,000 chemical substance index entries were generated. Of these chemical substance index entries, 368,000 corresponded to substances which were reported for the first time in 1975.

This paper addresses the computer-based algorithms that support the handling of chemical substance information. Since the methods used to represent information about chemical substances are critical to the nature of the algorithms used, a variety of chemical substance representation systems are presented, along with the various system processes necessary to handle computer-based files of chemical substance information. The algorithm developments that support these system processes are summarized, and sample algorithms are provided in the appendix to illustrate supporting system processes in areas of registration, substructure searching, and interconversions.

Lynch and others [4] provide an overview of principles and techniques for computer handling of information on chemical substances, and the characteristics of information handling systems utilizing these principles and techniques.

Representations of Chemical Substance Information

Chemical structure diagrams are two-dimensional visual descriptions of a chemical substance and provide an important medium for communications between chemists. Employing conventions for representing the three-dimensional structural features in the plane, these structure diagrams fall short of describing geometrical reality but they are the accepted way to describe chemical substances. Because structural diagrams are difficult to convey both orally and in written text, several other representation systems have been developed. Many of these chemical substance representation systems were developed prior to, but have been utilized in, computer-based chemical substance information handling systems. In addition, several representation systems more amenable to algorithmic computer processing have been developed.

For input, storage, manipulation, and output within computer-based systems, a representation of the chemical substance must be selected. The selection of a particular representation scheme for an information system is based on the size of the files to which it applies, the functions to be performed, the available hardware and software, and the desired balance between

manual and machine processes. The substance representation
system is critical to the nature of algorithms in computer-based
chemical substance information handling systems.
    Not all representations are of equivalent descriptive power.
Two important characteristics of a representation are unambiguity
and uniqueness. A representation is unique if, upon applying the
rules of the system to a chemical substance, only one representation can be derived. A representation is unambiguous if the
representation applies to only one chemical substance, although
there may be more than one possible representation for each chemical substance. For example, in Figure 1a, the systematic name
provides a unique, unambiguous representation. The molecular
formula, Figure 1b, is a unique but ambiguous representation;
unique because for any chemical substance there is only one
molecular formula, but ambiguous because isomers also have this
molecular formula. The arbitrarily numbered connection table,
Figure 1c, provides a non-unique, unambiguous representation.
The representation is unambiguous since it corresponds to one and
only one substance, but it is not unique because alternative
numberings of the connection table would result in different
representations for the same chemical substance (the connection
table representation is discussed in more detail below). In
addition to being categorized according to their uniqueness and
ambiguity, chemical substance representations commonly used within computer-based systems can be further classified as systematic
nomenclature, fragment codes, linear notations, connection tables,
and coordinate representations.

    Systematic Nomenclature. Systematic nomenclature provides
a unique, unambiguous representation of a chemical substance
by the application of a rigorous set of systematic nomenclature
rules. A representation of a chemical substance is constructed
by applying these nomenclature rules to combine terms which
describe the individual rings, chains, and functional groups
within the chemical substance. Chemical nomenclature provides
a representation which can be interpreted directly by the prac-
ticing chemist, is generally suitable for oral discourse, can be
used in a printed index, and is increasingly available in computer-readable files. Davis and Rush [5, Chapter 8] describe
the origin, development, and examples of systematic nomenclature
systems.
    Figure 2 provides an example of systematic nomenclature
utilizing the CHEMICAL ABSTRACTS NINTH COLLECTIVE INDEX Nomenclature Rules [6]. The systematic name in this example is
cyclohexanol, 2-chloro-. It is generated by (1) determining the
principal functional group, the OH group; (2) determining the
ring or chain to which it is directly attached, cyclohexane;
(3) naming the functional group and its attached ring, cyclohexanol; and (4) naming all other functional groups and skeletal
fragments, 2-chloro, where the locant 2 identifies the point of

attachment to the cyclohexane ring.

Fragment Codes. Fragment codes are a series of predefined
descriptors which are assigned to significant substructural
units, e.g., rings or functional groups. A given code is assigned to a chemical substance if the structural component occurs
within the chemical substance. Typically, fragment codes provide a unique, ambiguous description of a chemical substance.
With the introduction of punched-card systems, fragment code
systems became popular because of the simplicity of representation and the ease of the coding and searching operations. Since
fragment codes offer only a partial description of a chemical
substance based on predefined descriptors, there are situations
for which certain substructural components that were not initially anticipated and defined cannot be searched and situations
of extraneous retrievals of structures containing the needed
fragments but not in desired relationships. Although fragment
codes are valuable for subclassification of files, in the case
of large files, fragment codes are usually accompanied by other,
more complete representations. Figure 3 provides an example of
a fragment code representation utilizing the Ring Code System
[7], with codes corresponding to the card columns and punches
for the particular characteristic cited.

Linear Notation. Linear notation systems use a linear
string consisting of a set of symbols to represent complete topological descriptions of chemical substances. Each system has
symbols which represent atoms or groups of atoms, a syntax to
describe interconnections, and rules for ordering the symbols
to provide a unique and unambiguous representation of the topo-
logy of a chemical substance. After deriving a linear notation
by applying a set of ordering rules, linear notations are easy
to input and require no specialized input equipment. The
representation is very compact and the file structure is simple;
also linear notations can be utilized in printed indexes. Davis
and Rush [5, Chapter 9] provide general information on linear
notation systems and a more detailed discussion of the origin
and development of the IUPAC, Wiswesser, Hayward, and Skolnik
linear notation systems.

Figure 4 provides an example of a representation using
Wiswesser Line Notation. For this example, the Wiswesser Line
Notation is L6TJ AQ BG. The ring system is cited first and is
represented by L6TJ where L indicates the start of a carbo-
cyclic ring, 6 indicates a six-member ring, T indicates that the
ring is fully saturated, and J indicates the end of the ring
system. The substituents Cl and OH are represented by G and Q,
respectively, and their positions of attachment are identified
by the locants A and B. Since Q occurs later than G is the
defined collating sequence, Q is cited before G.

a.) Systematic Nomenclature: Benzene, 1,4-dichloro-

b.) Molecular Formula: $C_6H_4Cl_2$

c.) Connection Table:

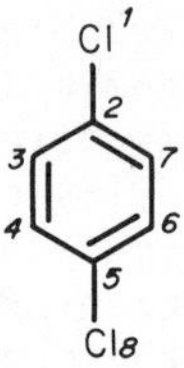

Atom No.	Element	Bonds	Connections
1	Cl	S	2
2	C	S,S,D	1,3,7
3	C	S,D	2,4
4	C	D,S	3,5
5	C	S,D,S	4,6,8
6	C	D,S	5,7
7	C	D,S	2,6
8	Cl	S	5

Figure 1. *Various representations of the chemical substance*

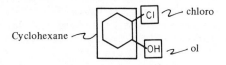

Cyclohexane

chloro

ol

Cyclohexanol, 2-chloro-

Figure 2. *Representation using systematic nomenclature*

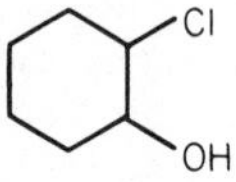

Code	Characteristic
2/12	One Isolated Ring
4/1	One 6-member Fully Saturated Carbocyclic Ring
17/1	Chlorine Present
18/1	One OH

Figure 3. *Representation using fragment codes*

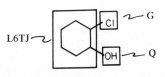

L6TJ

L6TJ AQ BG

Figure 4. *Representation using Wiswesser line notation*

Connection Tables. A structure diagram of a chemical substance can be viewed as a graph with the nodes corresponding to
the non-hydrogen atoms of the substance and the edges connecting
the nodes corresponding to the bonds of the substance. Given an
arbitrary numbering of the non-hydrogen nodes of the graph, the
connection table is a tabular description of the graph in which
each node is both listed in numerical order and is described by
the element symbol and the interconnections of each atom with
each other atom are explicitly described. Structural details
such as charge, abnormal valency, and isotopic mass can be
recorded with each atom. Beyond the atoms and bonds, the connection table introduces no concepts of chemical significance into
the representation. Consequently, connection tables can be
input by clerical staff with little training. Figure 5 provides
an example of a connection table. Since all interconnections are
cited twice, this form is called a redundant connection table.
By numbering the atoms of a structure such that once an atom has
been numbered, all un-numbered atoms directly connected to it
are numbered, and by citing only connections to lower-numbered
atoms, a more compact connection table can be derived. Figure
6 provides an example of a compact connection table. Since the
interconnection between Atom 7 and Atom 8 has not been cited,
these attachments, which complete the description of the interconnections of the structure, are cited in a field called the
ring closure list.

Dittmar, Stobaugh, and Watson [8] describe the connection
table utilized in the CAS Chemical Registry System. Lefkowitz
[9] describes a concise form of a connection table, called the
Mechanical Chemical Code, which does not explicitly identify the
bonds and has attributes of both a connection table and linear
notation.

The DARC code [10] resembles a connection table, since it
expresses or implies the nature of each atom and bond, but it is
generated in a concise, linear form. The description begins
with one atom which is chosen as the "focus" of the structure
and then proceeds outward, describing the "environment" of the
"focus."

Coordinate Representation. A coordinate representation of
a chemical substance is a recording of the atoms and bonds of
that substance with an indication of their relative position in
a plane. This coordinate representation provides a valuable form
to facilitate on-line, real-time manipulation of the structure
diagram and to store the diagram for subsequent composition in
journals, handbooks, and search output. Because this representation is difficult to manipulate, it is typically converted to
some other form for other information system functions. Farmer
and Schehr [11] describe the approaches and capabilities used at
CAS for representing and processing a coordinate form of structure diagrams.

Figure 7 graphically shows a coordinate representation of the chemical structure diagram. Every identifiable substructural unit has a node, symbolized by ◯ , associated with it. The node corresponding to the complete structure diagram is the root node and is the origin of the coordinate system. Every atom (including implied carbons) and bond of the structure has a leaf, symbolized by ▢ , associated with it. In the structure diagram, a leaf contains the characters for the element symbols or the line definitions for the bonds and their coordinates to indicate the position in the plane. Coordinate data for a leaf or node are relative to its parent node. Thus it is possible to change the coordinates of an entire subtree by changing the coordinates of the parent.

Processes

The ability to identify and collect all information about a particular chemical substance at one point is essential to computer-based chemical information handling systems. This eliminates the redundancy of work, e.g., in biological testing; it permits effective indexing of chemical substance information and it allows one to determine if a substance has been previously synthesized. The data base resulting from these processes can also be utilized for the identification of those substances with common structural characteristics. With the variety of chemical substance representation systems, the ability to interconvert between representations allows flexibility in performing system functions and permits the interchange of information among various chemical substance information handling systems. The system processes and algorithm development to support these processes are described below.

Registration. The registration of a chemical substance is the set of data management procedures which enables all information relating to a specific chemical substance to be linked together. The registration procedure is concerned with determining if a potentially new substance is equivalent to a substance already on file or if it is new, in which case the substance is added to the file.

The registration procedure used is determined by whether the structural representation is both unique and unambiguous. In systems without a unique and unambiguous representation of a chemical substance, the unique and unambiguous identification is accomplished through the registration processes. Initially, the file of substances is partitioned into small groups of substances on the basis of unique and ambiguous characteristics. For a potentially new substance, its unique and ambiguous characteristics are identified and final determination of whether the candidate substance is new is made by direct atom-by-atom structure comparison of the candidate with the subgroup of the

existing substances that have the same characteristics. The
selection of characteristics for the partitioning is obviously
critical, because the effectiveness of this registration technique is dependent on limiting the size of the subgroups. This
technique is called the isomer sort-registration technique.
Brown and others [12] describe the Merck, Sharp, and Dohme chemical structure information system which utilizes this approach.

In systems that use a unique and unambiguous representation,
determining if a potentially new substance is already on file
reduces to the comparison of the unique, unambiguous representation of candidate substance to the unqiue, unambiguous representation of the substances previously on file. With linear
notations, the unique, unambiguous representation is typically
achieved through manual encoding of the chemical substance.
Eakin [13] describes the chemical structure information system
at Imperial Chemical Industries Ltd., where registration is
based on Wiswesser Line Notation. For connection tables, the
unique, unambiguous representation is derived automatically,
i.e., a single, invariant numbering of the connection table is
algorithmically derived.

The algorithm used in the CAS Chemical Registry System to
generate a unique, unambiguous representation from an arbitrarily
numbered connection table [14] is described in a later section.
Dittmar, Stobaugh, and Watson [8] provide a description of the
general design of the CAS chemical structure information system
which utilizes a unique, unambiguous connection table.

Substructure Searching. Registration, as described in the
previous section, is a form of full-structure searching. Although the registration process is concerned with determining
if a complete structure existed previously within a collection,
the data base resulting from the registration processes can be
used for other purposes, in particular for substructure searching. Substructure searching is the identification of all substances within a file which contain a given partial structure.
Although substantial attention has been given to substructure
searching, several problems still remain, particularly in the
on-line substructure searching of large files, i.e., those that
contain more than a million substances.

With the variety of chemical substance representations, i.e.,
fragment codes, systematic nomenclature, linear notations, and
connection tables, a diversity of approaches and techniques are
used for substructure searching. Whereas unique, unambiguous
representations are essential for some registration processes,
it is important to note that this often cannot be used to
advantage in substructure searching. With connection tables,
there is no assurance that the atoms cited in the substructure
will be cited in the same order as the corresponding atoms in the
structure. With nomenclature or notation representation systems,
a substructural unit may be described by different terms or

symbols in the complete structure because of the context in which
the substructural unit appears.

Fragment code systems, devised to permit retrieval of a
chemical structure in a variety of ways, previously utilized
manually derived codes which were stored on and searched from
punched cards. With the development of computer techniques,
many of the early systems were expanded to permit the storage
and search of a wide variety of more complex codes. The fragments may correspond to general specific or structural features
and are often organized to allow searching at any level of
specificity. Search questions are stated in terms of the fragments used for representation and thus retrievals consist of
all substances containing the required fragments. Because the
addition of new structural features requires the re-analysis of
the previously processed file, attention has been given to the
automatic derivation of fragment codes from an unambiguous substance representation. The development of the Gremas fragment
code system at International Documentation in Chemistry [15] was
originally based on manually derived fragment codes but has
subsequently been expanded to generate the codes from connection
tables and topological descriptions that have been input by an
optical scanning device. Craig [16] describes the fragment codes
retrieval system used by Smith, Kline, & French Laboratories.

With the increasing availability of computer-readable files
of systematic nomenclature and capabilities for text searching,
attention has been given to the development of substructure
searching of files of systematic nomenclature using search terms
that are also systematic nomenclature terms. Fisanick and others
[17] describe an investigation into nomenclature-based substructure searching using techniques and search aids developed
at CAS.

Substructure searching based on linear notations can be
accomplished in both an automated and non-automated mode. Dyson
[18] describes a computer-produced permuted index that supports
the manual searching of the Dyson-IUPAC Linear Notation for substructural components. Computer-based substructure searching of
a linear notation involves examining the symbols of the linear
notation to determine if the substructural features exist.
Granito and Garfield [19] contrast substructure retrieval systems
based on fragment codes, connection tables, and linear notations.
In addition, they describe applications of Wiswesser Line Notation at the Institute for Scientific Information that support
substructure searching, registration, structure/property relationship studies, and display. Lynch and others [4, Chapter 5]
describe techniques and consideration for the computer-based
searching of linear notations. As with nomenclature substructure
searches, the success of a substructure search of linear notation depends directly on the ability of the questioner to
anticipate the environment of the required fragment in various
structures.

Depending on the sophistication needed, substructure searching can be accomplished with a variety of the representations of
a chemical substance. Some substructure searches can only be
adequately answered by a complete atom-by-atom and bond-by-bond
search for which a connection table, with its explicit description of full structural detail, is essential.

There are two approaches to the atom-by-atom substructure
search of a connection table: iterative atom-by-atom search [20]
and the Sussenguth set reduction technique [21]. Because each
of these specify alternative atoms and bonds and alternative
subgroups, there is virtually no limit to the degree of general-
ity or specificity of the search.

The iterative atom-by-atom search involves locating the
least commonly occurring atom in the substructure and searching
for each other atom of the substructure by path-tracing. When a
non-match is found, searching is continued by backing up to the
most recent branch point and proceeding along another path. This
iterative procedure is continued until the substructure is found
or the whole structure has been examined without finding the
substructure.

The Sussenguth set reduction technique involves partitioning the atoms of both the substructure and the structure based
on the atoms, bonds, and interconnections. The technique
involves generating subsets of atoms for the structure and the
subsets of atoms for the substructure, based on the elements,
bond values, and number of attachment. For example, all carbon
atoms would be in the same subset, all atoms with single bonds
attached would be in the same subset, etc. These subsets would
then be further partitioned by intersecting pairs of subsets --
e.g., all carbons with single bonds attached would be in a subset, all carbon with double bonds attached would be in the same
subset, etc. Additional subsets would then be generated using
the connections of each atom, and further partitioning would be
attempted. These processes for partitioning and generating sets
lead to one of the following situations: (1) a complete correspondence between each atom in the substructure and the structure, in which case the substructure is contained within the
structure; or (2) a non-correspondence between each atom of the
substructure and the structure, in which case the substructure
is not contained within the structure; or (3) a situation in
which no direct correspondence can be found, because either the
properties used to partition the atoms were not powerful enough
to distinguish between each atom or there is more than one
correspondence between the substructure and structure. In the
third case, the various alternatives for the correspondence
between substructure and structure must be tried, thus leading
to the correspondence or a contradiction.

Both of these approaches to substructure searching of a
connection table are extremely time-consuming, and it is usually
necessary for economic reasons to use some form of screening

Atom No.	Elements	Bonds	Connections
1	C	S,S	2,7
2	C	S,S	1,3
3	C	S,S	2,4
4	C	S,S	3,5
5	C	S,S,S	4,6,7
6	O	S	5
7	C	S,S,S	1,5,8
8	Cl	S	7

Figure 5.   Representation using connection table

Atom No.	Attachment	Element	Bond
1		C	
2	1	C	S
3	1	C	S
4	1	O	S
5	2	C	S
6	2	Cl	S
7	3	C	S
8	S	C	S
Ring Closure	7/8		S

*Figure 6.   Representation using compact connec-
tion table*

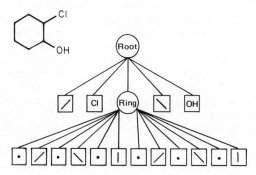

*Figure 7.   Coordinate representation of struc-
ture diagram*

Simple Pairs   C—C

Augmented Pairs
(where the connectivities
of the atoms are included)   2C—Cl

Bonded Pairs
(where the bond values
for attachments are included)

Figure 8.   Bond-centered fragments

system. In fact, it may be necessary to develop some form of a
screening system for large files, regardless of the representation system. Screening is the first stage of a substructure
search and is intended to inexpensively eliminate a large number
of structures which do not meet the requirements of a particular
substructure search question. Screens are characteristics which
can be identified in the substances in a file; they are similar
to fragment codes but usually consist of computer-generated data
of structural significance (elements, bonds, counts, small substructural units) rather than the nomenclature and function data
used in fragment code systems. After the screens are generated
for a particular substructure, the screen search is carried out
to select all structures which contain the characteristics
necessary for a particular substructure, thus minimizing the number of compounds requiring a detailed search.

In the selection of a screening system, the determination
of the set of structural characteristics to act as the screens is
a major problem. A proper balance must be established between
the cost of generating, storing, and searching the screens, and
insuring that the searches at the screen level achieve complete
recall. In addition, the structural characteristics selected
as screens should occur with a distribution as even as possible.
Because of the uneven distribution of structural characteristics,
this represents a significant problem.

Adamson and others [22] account for disparate frequencies of
characteristics in chemical structures by employing screens at
different levels of details. The screens for frequent characteristics are generated at a substantial level of detail whereas
less common characteristics are carried in more general terms.
For this approach, the set of screens are chosen on the basis of
the attributes and the size of the file. The screens thus selected are based on bond-centered fragments with three different
levels of detail as illustrated in Figure 8. The most commonly
occurring pairs of atoms in the file are included as screens
among the simple pairs. For a sample file of 30,000 structures
from the CAS Chemical Registry System, 18 simple pairs were
included. The most frequently occurring simple pairs were
included as augmented pairs screens. For the particular file
studied, the augmented pairs all involved carbon attached to
carbon (CC), carbon attached to nitrogen (CN), or carbon attached
to oxygen (CO). The most frequently occurring augmented pairs
were included as bonded pair screens; again these involved only
CC, CN, or CO. The total set of screens consisted of (1) the
number of common structural features, e.g., the number of carbon
atoms, the number of atoms with connectivity equal to 3, or the
number of double-chain bonds; (2) bits to indicate the presence
or absence of various atoms; (3) bits to indicate the presence
or absence of the 18 most common simple pairs of atoms; (4) bits
to indicate the presence or absence of the augmented pairs; (5)
bits to indicate the presence or absence of bonded pairs, and

(6) bits to indicate the presence or absence of various ring
systems. A description of the algorithm that generates these
screens is provided in a later section.

To achieve an even distribution of screens and a wider
variation in fragment selection, Feldman and Hodes [23] have
developed a screen generation procedure for use in the chemical
structure search system at the Walter Reed Army Institute of
Research. The screens selected are based on frequency statistics
from a sample of the total base. The process involves "growing"
fragments for each structure from a subset of their file by
starting with each atom and then adding single atoms at each
iteration to the fragments generated during the previous iteration. This process would generate all possible fragments. To
keep the number of fragments at a reasonable number, an elimination rule based on the frequency of occurrence of that fragment
within the sample file is applied. This rule determines which
fragments are to be eliminated (those which occur at a frequency
of less than 0.1%), and which fragments are to be passed on to
the next iteration (those which occur at a frequency of greater
than 1%), where they will "grow" further. In addition, a
heuristic procedure based on earlier operational experience was
used to "prune" a large number of fragments which were chemically
insignificant. The fragments obtained at the completion of this
iterative process were then used as screens.

Interconversion. With the variety of representations, the
approach taken in selection of a chemical substance representation has not been to select one representation to handle a full
range of functions, but rather, through automatic interconversion, to utilize the representation which best solves a particular problem or meets a particular set of processing requirements for a given information system. In addition to providing
this internal flexibility, automatic interconversion permits
interchange of information among systems using various structure
representations. Granito [24] discusses the needs and status of
interconversions among chemical substance information systems.
Campey, Hyde, and Jackson [25] illustrate a chemical structure
information system which uses a variety of representations.

Substantial attention and progress has been made in the
development of procedures to effect conversion between chemical
substance representations. Zamora and Davis [26] describe an
algorithm to convert a coordinate representation of a chemical
substance (derived from input by a chemical typewriter) to a
connection table. An approach for interactive input of a
structure diagram and conversion of this representation to a
connection table suitable for substructure searching is discussed
by Feldmann [27]. The conversion of systematic nomenclature to
connection tables offers a powerful editing tool as well as a
potential mechanism for conversion of name files to connection
tables; this type of conversion is described by Vander Stouw [28].

Programs now exist to convert Wiswesser Line Notation [29], Hayward [30], and IUPAC [18] linear notations to connection tables. Because fragment codes alone do not provide the complete description of all structural detail, conversion to other representations is typically not possible.

The conversion from a connection table to other unambiguous representations is substantially more difficult. The connection table is the least structured representation and incorporates no concepts of chemical significance beyond the list of atoms, bonds, and connections. A complex set of rules must be applied in order to derive nomenclature and linear notation representations. To translate from these more structured representations to a connection table requires primarily the interpretation of symbols and syntax. The opposite conversion, from the connection table to linear notation, nomenclature, or coordinate representation first requires the detailed analysis of the connection table to identify appropriate substructural units. The complex ordering rules of the nomenclature or notation system or the esthetic rules for graphic display are then applied to derive the desired representation.

Ebe and Zamora [31], building on algorithms that generate Wiswesser Line Notation for ring systems from a connection table [32], have developed procedures to employ these interconversions for editing Wiswesser Line Notations for complex ring systems. Farrell, Chauvenet, and Koniver [33] describe procedures for generating Wiswesser Line Notation from connection tables and Lefkovitz [9] describes the derivation of Mechanical Chemical Code, a concise form of a connection table from the CAS connection table. Programs have also been developed to derive a DARC code from both connection tables and linear notations. Algorithms for generation of systematic nomenclature from a connection table are currently being developed by CAS.

Because the structure diagram is a desirable form of output from an automated chemical structure information handling system, several algorithms have been developed to generate a coordinate representation from a connection table [34 and 35]. However, most structure display systems were developed for a chemical typewriter or line printer, and the physical characteristics of these devices restrict the complexity of structures to be displayed. An algorithm for a general Cartesian coordinate system, which produces structure diagrams of high graphical quality from a connection table representation, has been developed and utilized at CAS and is described by Dittmar and Mockus [36]. In a later section, an example is provided to illustrate features of this algorithm.

Related Continuing Developments

A variety of algorithms for the computer handling of chemical structure information have been described. The techniques for

representation and processing have become established, and, as
indicated by the existence of effective operational systems
[4, Chapters 8 and 9] and some algorithms presented earlier,
practical solutions exist for many of the problems in the handling of chemical structures.

Several of the general graph theory problems are presently
unsolved. An example is subgraph isomorphism: given two graphs,
G_1 and G_2, is G_1 isomorphic to a subgraph of G_2? It is con-
jectured that no algorithm for solving it in polynomial time
exists, i.e., all known algorithms have at least an exponential
growth rate based on the number of vertices for some subset of
graphs. Another example is general graph isomorphism: given
two graphs, G_1 and G_2, is G_1 isomorphic to G_2? This problem is
also unsolved and is a special case of the subgraph isomorphism
problem [37]. For various classes of graphs, as in the case of
planar graphs [38], isomorphism algorithms have been found.
Sanders [39] demonstrates that the algorithmic generation of
Wiswesser Line Notation is not polynomial bounded. As illustrated earlier, good heuristic procedures have been established
to provide solutions to isomorphism problems for the graphs
corresponding to chemical structures. However, the general
graph theory problems remain and are receiving continued attention.

Algorithms that process structural data of chemical substances are being developed for many areas. For example,
structure/property correlation [40] utilizes a chemical substance
data base to provide a correlation between biological properties
and structural features of chemical substances. Reactants and
products of chemical reactions can be analyzed to provide
retrieval of information about partial structures that
characterize the reaction [41]. Among the computer programs
that have been developed for utilizing chemical structure information are molecular modeling programs [42], aimed at using the
computer to generate actual three-dimensional descriptions of
chemical substances, and organic synthesis programs [43], which
predict by computer the design of possible synthetic routes to
a given target substance.

APPENDIX

Sample Algorithms

Illustrative sample algorithms that support system processes
in areas of registration, substructure searching, and automatic
interconversion are provided below.

Algorithm I - Registration - Canonicalization of Connection
Tables. A connection table for a chemical substance with n
atoms can be numbered in as many as n! different ways. The
problem of generating a canonical form involves selecting a

single and invariant numbering of the connection table. An approach would be to generate all n! representations, sort them alphabetically, and then select the one which compares low. Except for very small n, this procedure is obviously not feasible. The approach presented below is a variation of this procedure, and limits the number of representations that must be generated by establishing a partial order of atoms, restricting the numbering permitted, and saving the results of the path tracing.

Given an arbitrarily numbered connection table representation of a structure with n non-hydrogen atoms, the unique numbering is obtained as follows:

1. Assign Stage 1 connectivity values to each atom based on the number of attachments to the atoms.

2. Assign Stage 2 connectivity values to each atom by summing the Stage 1 connectivity values for the attached atoms.

3. Given the Stage i connectivity values for each atom, assign the Stage i+1 connectivity values by summing the Stage i connectivity values for the attached atoms.

4. Calculate the number of distinct connectivity values at the Stage i and Stage i+1.

5. If the number of distinct connectivity values at the Stage i+1 is greater than Stage i, go to step 3.

6. Otherwise, the final connectivity values are the Stage i values.

7. Select the atom with the highest connectivity value and designate that atom as Number 1.

8. Since Steps 1-6 provide only a partial order of the atoms, note all other atoms with same connectivity value.

9. Atoms connected to Atom 1 are assigned 2, 3, etc. based on decreasing connectivity values. If a choice is arbitrary (where the atoms have the same connectivity value) note the pairs of atoms involved in the arbitrary choice.

10. The unnumbered atoms attached to Atom 2 are numbered based on decreasing connectivity values. Again, note pairs of atoms where the choice was arbitrary.

11. This procedure is followed until all atoms have been numbered.

12. Build and retain the compact connection table based on this numbering.

13. Back up to the highest numbered atom for which the choice was arbitrary. If there are no remaining atoms where the choice was arbitrary, the process is complete and the retained connection table is the unique representation.

14. Select the other atom from the pair involved in the arbitrary choice and renumber the atoms of the structure from that atom to the last atom.

15. Build a new compact connection table.

16. Compare the newly generated compact connection table to the retained compact connection table.

17. If the new connection table is alphabetically less than the retained table, replace the retained table with the new table, and go to Step 13.

18. Otherwise, go to Step 13.

Figure 9 illustrates the steps in the algorithm for generating the unique connection table. Figure 9a illustrates the Stage 1 connectivity values which are the number of attachments, and Figure 9b illustrates the Stage 2 connectivity values which are obtained by summing the Stage 1 connectivity values for the attached atoms. At Stage 2, the number of distinct values is 4. The Stage 3 connectivity values are obtained by summing the Stage 2 connectivity values for the attached atoms, as illustrated in Figure 9c. Since in Stage 3 the number of distinct values is 6, which is greater than the Stage 2 value of 4, the iterative process is continued. Figure 9d illustrates the Stage 4 connectivity value calculations. Since the number of distinct values at Stage 4 is equal to that at Stage 3, the final connectivity values assigned are those calculated at Stage 3.

Figure 9e illustrates the initial numbering and the compact connection table using the Stage 3 connectivity values. The atom with connectivity value of 13 is assigned Number 1. The atoms attached to Atom 1 are numbered 2, 3, and 4 based on decreasing connectivity values. The arbitrary choice between 3 and 4 is noted.

The unnumbered attachment to Number 2 is assigned Number 5. The unnumbered attachments to Atoms 3, 4, 5 are numbered. Based on this numbering, the initial connection table is constructed

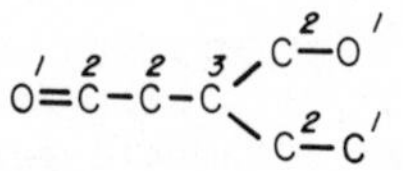

$$O \overset{1}{=} \overset{2}{C} - \overset{2}{C} - \overset{3}{C} \underset{\underset{C \overset{2}{=} C'}{\diagdown}}{\overset{\overset{C \overset{2}{=} O'}{\diagup}}{}}$$

a.) Stage 1 Connectivity Values = $\{1,2,3\}$.
   No. of Distinct Values = 3.

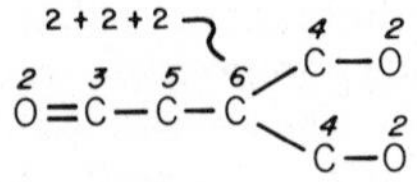

$$\overset{2+2+2}{\underset{0}{\overset{2}{=}} \overset{3}{C} - \overset{5}{C} - \overset{6}{C}} \underset{\underset{C - O}{\overset{4}{\diagdown} \overset{2}{}}}{\overset{\overset{C - O}{\overset{4}{\diagup} \overset{2}{}}}{}}$$

b.) Stage 2 Connectivity Values = $\{2,3,4,5,6\}$.
   No. of Distinct Values = 5.

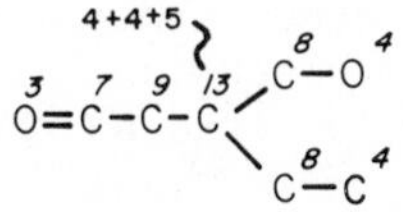

$$\overset{4+4+5}{\underset{0}{\overset{3}{=}} \overset{7}{C} - \overset{9}{C} - \overset{13}{C}} \underset{\underset{C - C}{\overset{8}{\diagdown} \overset{4}{}}}{\overset{\overset{C - O}{\overset{8}{\diagup} \overset{4}{}}}{}}$$

c.) Stage 3 Connectivity Values = $\{3,4,7,8,9,13\}$.
   No. of Distinct Values = 6.

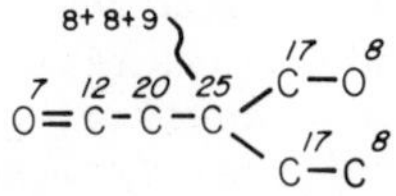

$$\overset{8+8+9}{\underset{0}{\overset{7}{=}} \overset{12}{C} - \overset{20}{C} - \overset{25}{C}} \underset{\underset{C - C}{\overset{17}{\diagdown} \overset{8}{}}}{\overset{\overset{C - O}{\overset{17}{\diagup} \overset{8}{}}}{}}$$

d.) Stage 4 Connectivity Values = $\{7,8,12,17,20,25\}$.
   No. of Distinct Values = 6.

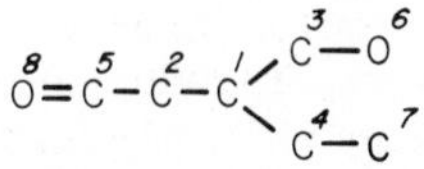

$$\overset{8}{0} \overset{5}{=} \overset{2}{C} - \overset{1}{C} - C \underset{\underset{C - C}{\overset{4}{\diagdown} \overset{7}{}}}{\overset{\overset{C - O}{\overset{3}{\diagup} \overset{6}{}}}{}}$$

Atom No.	1	2	3	4	5	6	7	8
Attachments	1	1	1	2	3	4	5	
Elements	C	C	C	C	C	O	C	O
Bonds	S	S	S	S	S	S	D	

e.) Initially Numbered Connection Table.
   Arbitrary Choice Between 3 and 4.

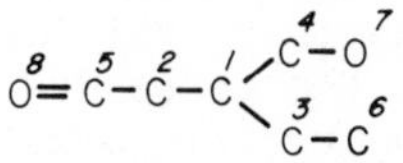

$$\overset{8}{0} \overset{5}{=} \overset{2}{C} - \overset{1}{C} - C \underset{\underset{C - C}{\overset{3}{\diagdown} \overset{6}{}}}{\overset{\overset{C - O}{\overset{4}{\diagup} \overset{7}{}}}{}}$$

Atom No.	1	2	3	4	5	6	7	8
Attachments	1	1	1	2	3	4	5	
Elements	C	C	C	C	C	C	O	O
Bonds	S	S	S	S	S	S	D	

f.) Alternately Numbered Connection Table.

Figure 9.   Generation of a unique, unambiguous
connection table

and retained, as shown in Figure 9e.

Backing up to the highest atom marked as an arbitrary choice, Atom 3, the other alternative, is tried and the representation is renumbered from Atom 3. Figure 9f illustrates the numbering and compact connection table resulting from this alternative. The connection table generated is alphabetically compared to the retained connection table. The attachment lists of the retained and newly generated connected table are compared and they are equal. The atom list of the newly generated connection table is compared and is lower than the retained connection table, because C in Position 6 of the newly generated table is lower than 0 in the retained connection table. Therefore, the newly generated connection table is retained.

Since there are no other atoms noted as involving an arbitrary choice, the retained table is the single invariant representation which is selected as the representation for this substance.

The number of alternate numberings which must be attempted is dependent on the numbers of atoms which have attachments with equal connectivity values. All of these various alternate numbering combinations must be attempted. Consequently, the algorithm does not provide a practical solution to the general graph isomorphism problem. However, because the graphs corresponding to chemical structures typically have connectivities of 1, 2, 3, or 4, the algorithm does provide a practical way to uniquely label virtually all graphs corresponding to a chemical structure.

This algorithm is implemented on an IBM 370/168. As part of routine production at CAS, 13,000 substances per week are uniquely numbered through this algorithm at an average processing rate of 1000 structures per minute of CPU time. Since there are some highly symmetrical structures which would require a substantial number of iterations, the algorithm is implemented to stop after three CPU seconds and use a registration approach based on a non-unique representation. For the 677,000 substances processed in 1975, 990 substances could not be uniquely labeled within the three seconds. Ferrocene, shown in Figure 10, is an example of a structure which would require 10! or 3,628,800 iterations. For substances of this type which cannot be uniquely labeled within the three CPU second time limit, an isomer-sort registration technique is utilized to complete the registration processes without human intervention.

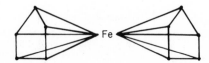

Figure 10.   Ferrocene

Algorithm II - Substructure Search - Screen Generation.   In an earlier section, bond-centered screens for substructure search are described.   Below is an algorithm for generating these screens.   Given the connection table representation of a chemical substance, the algorithm for the generation of the bond-centered screens consists of the following steps:

1. Construct the set of counts of atoms, bonds, and connections and set the appropriate atom and ring system bits.

2. Select the first/next pair of atoms.

3. If the pair is CC, CN, or CO, determine if it is one of the bonded pairs.   If not, go to Step 8.

4. If it is one of the bonded pairs, go to Step 10.

5. If it is not one of the bonded pairs, determine if it is one of the augmented atom pairs.

6. If it is one of the augmented atom pairs, go to Step 11.

7. If it is not one of the augmented atom pairs, go to Step 12.

8. If it is one of the simple pairs, go to Step 12.

9. If it is not one of the simple pairs, set exception pair bits and go to Step 13.

10. Set appropriate bonded pair bits.

11. Set appropriate augmented pair bits.

12. Set appropriate simple pair bits.

13. If this is not the last pair, go to Step 2.

14. If this is the last pair, the process is complete.

Algorithm III - Interconversion - Connection Table to Structure Diagram.   This algorithm has as input the connection table representation of a chemical substance and an authority file containing a coordinate representation of all unique ring system shapes for all ring systems; an example of input for one chemical substance is shown in Figure 11a.   The manually built file of coordinate representations for the ring system shapes eliminates many of the problems associated with assigning coordinates to ring systems.   This file at CAS contains 15,000 ring system shapes which represent the ring shapes for virtually all ring systems occurring with 3.5 X 10^6 distinct substances in the CAS Chemical Registry System. The examples below illustrate features of this algorithm.

The algorithm partitions the connection table into three groups: ring systems, the largest connected substructural units in which all edges are in a cycle; chains, linear acyclic strings with one terminal atom; and links, linear acyclic strings without any terminal atoms. The algorithm substitutes commonly recognized shortcut symbols for various groups of atoms, e.g., Me for the methyl group and Ph for the benzene ring. Figure 11b illustrates these processes.

The most central ring system is identified, its pre-stored ring shape is retrieved, and the nodes and the bonds of the ring system are mapped into the ring shape. The atom characters and bond vectors are calculated based on the coordinates of the ring shape, shown in Figure 11c. If there are no ring systems, the most central acyclic atom is used as the starting point.

With the most central ring system as the base structure, the direction, bond angle, and bond length are determined, first for the attached links and then for the chains attached to the ring systems. For links, the direction is away from the base structure, and is horizontal or vertical based on the angle nearest to the bisecting angle of the ring perimeter. For chains, the direction is away from the base structure and bisects the ring perimeter angle. A standard length bond is used. Figure 11d illustrates these processes.

For links and chains attached to the base structure, the coordinates of the atoms and bonds of the component are determined. The coordinates of the first atom attached to the ring system are determined. Coordinates for the next atom are above, below, to the right, or to the left, and they are determined based on the drawing direction. Horizontal single bonds are drawn implicitly; all other bonds are drawn explicitly. All atoms in the link or chain are placed similarly. When the coordinates of all links and chains are determined, the link

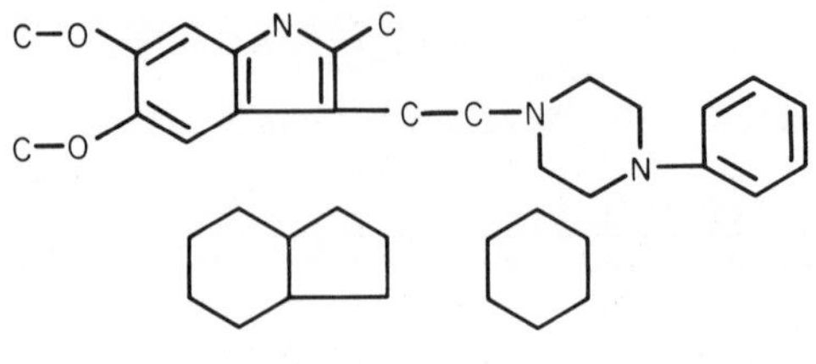

a) Connection Table for Substance and Coordinate Representation for Ring Shapes.

Figure 11. Generation of a coordinate representation from a connection table (continued on facing page)

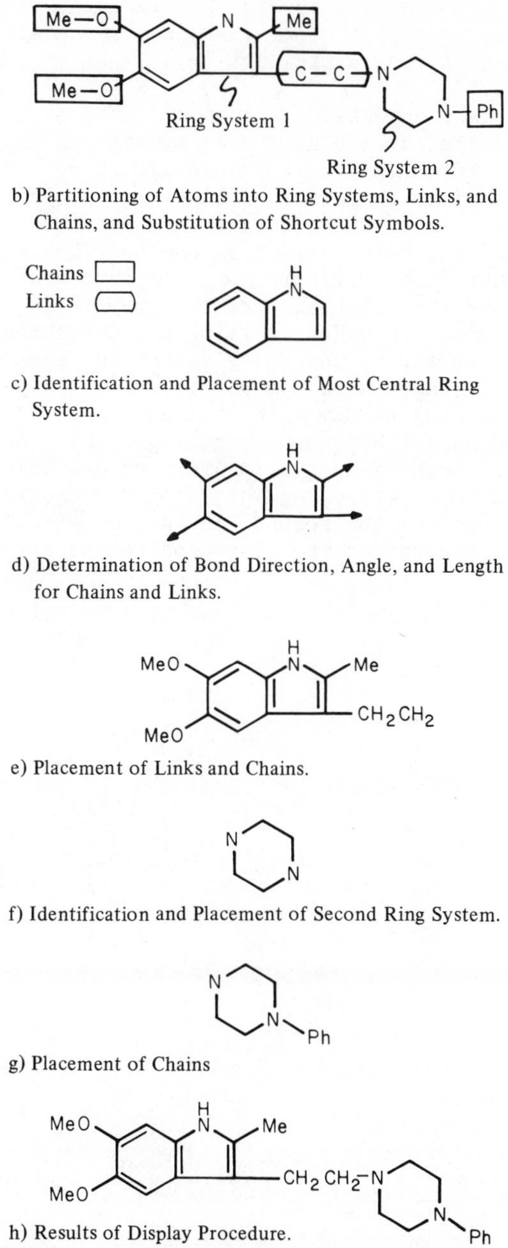
Me—O
Me—O
N
Me
C—C
N
N
Ph
Ring System 1
Ring System 2
b) Partitioning of Atoms into Ring Systems, Links, and Chains, and Substitution of Shortcut Symbols.
Chains
Links
H
N
c) Identification and Placement of Most Central Ring System.
H
N
d) Determination of Bond Direction, Angle, and Length for Chains and Links.
MeO
H
N
Me
CH2CH2
MeO
e) Placement of Links and Chains.
N
N
f) Identification and Placement of Second Ring System.
N
N
Ph
g) Placement of Chains
MeO
H
N
Me
CH2 CH2–N
MeO
N
Ph
h) Results of Display Procedure.

Figure 11. Generation of a coordinate representation from a connection table (continued from facing page)

or chain is positioned relative to the ring system, as illustrated in Figure 11e. All other links and the chains attached
to the most central ring system are positioned in a similar
manner.

When all links and chains attached to the most central
ring system are placed, the next ring system and its ring shape
are retrieved. (Note that in this example there are no links
or chains attached to the attached links and chains.) The atoms
and bonds are mapped into the ring shape, and the atom characters
and bond vectors are calculated from the coordinates of the ring
system, as illustrated by Figure 11f. The orientation of each
ring system after the first must reflect how it is attached to
the base structure. In order to allow for attaching it to the
base structure, it may be necessary to reflect the ring system
about the x-axis, the y-axis, or both.

If a second ring system with attachments is present, the
direction, bond angle, and bond length for chains and links
attached to the second ring system are then determined, as
shown in Figure 11g. Following this, the coordinates of links
and chains attached to the second ring system are attached. If
attachments are present on the links and chains attached to the

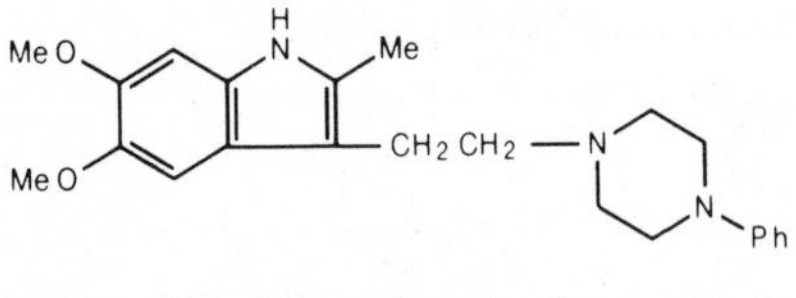

*Figure 12. Example of photocomposer
output*

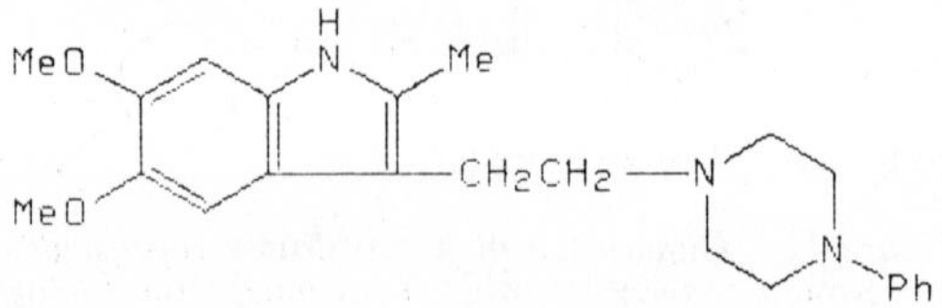

Figure 13. Example of electrostatic printer output

second ring system, they would be positioned at this point. The second ring system with its attachments is then attached to the base structure. Since all components of the substance have been processed, the display is complete; that is, a coordinate representation has been derived. The results of this process are illustrated in Figure 11h.

Throughout this process, as each component is added to the base structure, it is tested for overlap. If overlap is detected, it is resolved by extending the bond length and/or adjusting the bond angle. Since this algorithm uses the coordinate representation described earlier, movement of each component to be added requires updating the coordinates of the node associated with that component rather than the coordinates of each atom involved.

This algorithm produces highly acceptable results. With initial implementation, considerations for handling special cases of substances, e.g., coordination compounds, polymers, and incompletely defined structures, were deferred. The algorithm will generate images for many of these structures but acceptability is dependent on use. It is estimated that the current version of the algorithm will generate a highly acceptable (by CAS internal drawing standards) coordinate representation for 85% of the 3.5×10^6 unique substance in the CAS Chemical Registry System. The algorithm requires 266K bytes of main storage for executable instructions and processes 8 substances per CPU second on an IBM 370/168.

Within the CAS Composition Facility, the device-independent coordinate representation generated by this algorithm can be converted to the device-specific coordinates of the Autologic APS-4 photocomposer for high graphical quality output -- illustrated by Figure 12 -- or to the Varian Status 21 electrostatic printer for low cost worksheet production -- illustrated by Figure 13.

Literature Cited

1. "Toward a Modern Secondary Information System for Chemistry and Chemical Engineering," Chemical & Engineering News, 53, 30 (16 June 1975).
2. Lynch, Michael F., Computer-Based Information Services in Science and Technology, Peter Poregrinus Ltd., Herts, England, 1974.
3. ADI/ASIS, Cuadra, Carlos A. (ed.), Annual Review of Information Science and Technology, 1-10, Wiley/Interscience, (1966-1974).
4. Lynch, Michael F., Judith M. Harrison, William G. Town, and Janet E. Ash, Computer Handling of Chemical Structure Information, American Elseview Publishing Company, Inc., New York, N.Y., 1971.

5. Davis, Charles H. and James E. Rush, "Information Retrieval
    and Documentation in Chemistry," in Contributions in
    Librarianship and Information Science, Number 8, Greenwood
    Press, Westport, Connecticut, 1974.
6. Donaldson, N., W. H. Powell, R. J. Rowlett, Jr., R. W.
    White, and K. V. Yorka, "CHEMICAL ABSTRACTS Index Names
    for Chemical Substances in the Ninth Collective Period (1972-
    1976)," Journal of Chemical Documentation, 14(1), 3-14 (1974).
7. Oatfield, Harold, "The ARCS System: Ringdoc as Used with
    a Computer," Journal of Chemical Documentation, 7(1), 37-43
    (1967).
8. Dittmar, P. G., R. E. Stobaugh, and C. E. Watson, "The
    Chemical Abstracts Service Chemical Registry System. I.
    General Design," Journal of Chemical Information and Computer
    Sciences, 16(2), 111-121 (1976).
9. Lefkovitz, David, "A Chemical Notation and Code for Computer
    Manipulation," Journal of Chemical Documentation, 1(4), 186-
    192 (1967).
10. Dubois, J. E., "DARC System in Chemistry," in Computer Repre-
    sentations and Manipulation of Chemical Information, ed.
    W. T. Wipke and others, John Wiley & Sons, Inc., New York,
    N.Y., 1974.
11. Farmer, N. A. and J. C. Schehr, "A Computer-Based System for
    Input, Storage and Photocomposition of Graphical Data,"
    Proceedings of the ACM, Vol. 2, 563-570 (November 1974).
12. Brown, H. D., Marianne Costlow, Frank A. Cutler, Albert N.
    DeMott, Walter B. Gall, David P. Jacobus, and Charles J.
    Miller, "The Computer-Based Chemical Structure Information
    System of Merck, Sharp and Dohme Research Laboratories,"
    Journal of Chemical Information and Computer Sciences, 16(1),
    5-10 (1976).
13. Eakin, Diane R., "The ICI CROSSBOW System," in Chemical
    Information Systems, ed. J. E. Ash and E. Hyde, John Wiley
    & Sons, Inc., New York, N.Y., 1975.
14. Morgan, H. L., "The Generation of a Unique Machine Descrip-
    tion for Chemical Structures -- A Technique Developed at
    Chemical Abstracts Service," Journal of Chemical Documenta-
    tion, 5(2), 107-113 (1965).
15. Fugmann, R., "The IDC System," in Chemical Information
    Systems, ed. J. E. Ash and E. Hyde, John Wiley & Sons, Inc.,
    New York, N.Y., 1975.
16. Craig, P. N. and N. M. Ebert, "Eleven Years of Structure
    Retrieval Using the SK&F Fragment Codes," Journal of Chemical
    Documentation, 9(3), 141-146 (1969).
17. Fisanick, W., L. D. Mitchell, J. A. Scott, and G. G. Vander
    Stouw, "Substructure Searching of Computer-Readable Chemical
    Abstracts Service Ninth Collective Index Nomenclature Files,"
    Journal of Chemical Information and Computer Sciences, 15(2),
    73-84 (1975).

18. Dyson, G. M., "The Dyson-IUPAC Notation," in Chemical Infor-
    mation Systems, ed. J. E. Ash and E. Hyde, John Wiley &
    Sons, Inc., New York, N.Y., 1975.
19. Granito, Charles E. and Eugene Garfield, "Substructure
    Search and Correlation in the Management of Chemical Infor-
    mation," Naturwissenscheften, 60(4), 189-197 (1973).
20. Ray, L. C. and R. A. Kirsch, "Finding Chemical Records by
    Digital Computers," Science, 126, 814-819, (1957).
21. Sussenguth, Edward H., Jr., "A Graph-Theoretic Algorithm for
    Matching Chemical Structures," Journal of Chemical Docu-
    mentation, 5(1), 36-43 (1965).
22. Adamson, George W., Jeanne Cowell, Michael F. Lynch, Alice
    H. W. McLure, William G. Town, and Margaret A. Yapp,
    "Strategic Considerations in the Design of a Screening System
    for Substructure Searches of Chemical Structure Files,"
    Journal of Chemical Documentation, 13(3), 153-157 (1973).
23. Feldman, Alfred, and Louis Hodes, "An Efficient Design for
    Chemical Structure Searching I, the Screens," Journal of
    Chemical Information and Computer Sciences, 15(3), 147-151
    (1975).
24. Granito, Charles E., "CHEMTRAN and the Interconversion of
    Chemical Substructure Search Systems," Journal of Chemical
    Documentation, 13(2), 72-74 (1973).
25. Campey, Lucille H., E. Hyde, and Angela R. H. Jackson,
    "Interconversion of Chemical Structure Systems," Chemistry
    in Britain, 6(10), 427-430 (1970).
26. Zamora, Antonio, and David L. Dayton, "The Chemical
    Abstracts Service Chemical Registry System. V. Structure
    Input and Editing," to be published in the August 1976 issue
    of Journal of Chemical Information and Computer Sciences.
27. Feldman, R. J., "Interactive Graphic Chemical Structure
    Searching," in Computer Representation and Manipulation of
    Chemical Information, ed. W. T. Wipke, John Wiley, N.Y., 1974.
28. Vander Stouw, G. G., P. M. Elliott, and A. C. Isenberg,
    "Automated Conversion of Chemical Substance Names to Atom-
    Bond Connection Tables," Journal of Chemical Documentation,
    14(4), 185-193 (1974).
29. Hyde, E., F. W. Matthews, Lucille H. Thompson, and W. J.
    Wiswesser, "Conversion of Wiswesser Notation to a Connecti-
    vity Matrix for Organic Compounds," Journal of Chemical
    Documentation, 7(4), 200-203 (1967).
30. Tauber, S. J., S. J. Fraction, and H. W. Hayward, Chemical
    Structures as Information-Representations, Transformations,
    and Calculations, Spartan Books, Washington, D. C., 1965.
31. Ebe, Tommy, and Antonio Zamora, "PATHFINDER II, A Computer
    Program That Generates Wiswesser Line Notations for Poly-
    cyclic Structures," Journal of Chemical Information and
    Computer Sciences, 16(1), 36-39 (1976).

32. Bowman, C. M., F. A. Landee, N. W. Lee, and M. H. Reslock, "A Chemically Oriented Information Storage and Retrieval System II. Computer Generation of the Wiswesser Notation of Complex Polycyclic Structures," Journal of Chemical Documentation, 8(3), 133-138 (1968).

33. Farrell, C. D., A. R. Chauvenet, and D. A. Koniver, "Computer Generation of Wiswesser Line Notation," Journal of Chemical Documentation, 11(1), 52-59 (1971).

34. Rogers, M. A. T., "CROSSBOW," presented at the 158th National Meeting of the American Chemical Society, New York, N.Y., September 1969.

35. Zimmerman, B. L., Computer-Generated Chemical Structural Formulas with Standard Ring Orientations, Ph. D. Dissertation, University of Pennsylvania, Philadelphia, Pennsylvania, 1971.

36. Dittmar, Paul G. and Joseph Mockus, "An Algorithmic Computer Graphics Program for Generating Chemical Structure Diagrams," submitted to Journal of Chemical Information and Computer Science.

37. Barrow, H. G. and R. M. Burstall, "Subgraph Isomorphism, Matching Relational Structures, and Maximal Cliques," Information Processing Letters, 4(4), 83-84 (January 1976).

38. Hopcroft, J. E. and R. E. Tarjan, "Isomorphism of Planar Graphs," in Complexity of Computer Computations, ed. Raymond E. Miller and James W. Thatcher, Plenum Press, New York, 1972.

39. Sanders, Alton F., "Graph Theoretical Constraints on Linearization Algorithms for Canonical Chemical Nomenclature," presented at the 169th National Meeting of the American Chemical Society, Philadelphia, April, 1975.

40. Jurs, P. C. and T. L. Isenhour, Chemical Applications of Pattern Recognition, John Wiley & Sons, Inc., New York, N.Y., 1975.

41. Valls, J., "Chemical Reaction Indexing," in Chemical Information Systems, ed. J. E. Ash and E. Hyde, John Wiley & Sons, Inc., New York, N.Y., 1975.

42. Marshall, G. R., H. E. Bosshard, and R. A. Ellis, "Computer Handling of Chemical Structures: Applications in Crystallography, Conformational Analysis, and Drug Design," in Computer Representation and Manipulation of Chemical Information, ed. W. T. Wipke and others, John Wiley & Sons, Inc., New York, N.Y., 1974.

43. Bersohn, M., and A. Esack, "Computers and Organic Synthesis," Chemical Reviews, 76(2), 269-282 (1976).

INDEX